Religion: The Enemy of Reason

How Religion Feeds Superstition, Weakens Science, and Silences Questioning

---

Dr Prabhat Das

MD, PhD, FACP

BOOK 1

Religion: The Holy Poison

Hardcover ISBN: 979-8-951067-19-7

Paperback ISBN: 979-8-951067-19-7

eBook ISBN: 979-8-951067-50-0

Printed in the United States of America

TABLE OF CONTENTS:

Introduction: Don't Believe Me. Follow the Evidence.

The Question That Changes the Room

A child asks, "How do we know God is real?"

For a moment, the room becomes quiet.

An adult may answer, "You must have faith." Another may say that God is beyond human understanding. Someone else may smile, change the subject, or remind the child that good children should not question sacred things.

The response is not always harsh. It often comes from loving parents or grandparents who believe they are protecting the child. They are passing forward what they themselves were taught.

Still, the child notices something important.

Questions are welcomed when they concern nature, school, machines, or ordinary life. But when the subject becomes God, scripture, prayer, miracles, or religious tradition, curiosity may suddenly look like disrespect.

The child learns that some questions are safe and others are dangerous.

That lesson may remain for life.

Do not question God.

Do not doubt scripture.

Do not challenge tradition.

Slowly, curiosity becomes uncomfortable. Doubt begins to carry guilt. Obedience begins to look like virtue.

This is one of religion's earliest injuries to reason.

Religion does not merely provide answers. It often decides which answers may never be questioned.

That is where this book begins.

Why Reason Matters

Reason is sometimes described as cold, arrogant, or hostile to faith.

It is none of those things.

Reason is the habit of asking whether a claim is supported, whether an explanation fits the facts, whether an authority may be mistaken, and whether a tradition is causing harm.

It is one of humanity's greatest protections.

Without it, disease can become punishment, poverty can become fate, caste can become sacred order, and obedience can become holiness. A ritual may be accepted because it is old. A miracle may be believed because the story is moving. A child may be frightened because an adult speaks with confidence.

The problem is not that religious people are unintelligent.

Many believers are thoughtful, educated, compassionate, and professionally accomplished. They are physicians, engineers, professors, scientists, lawyers, teachers, and public leaders. They use evidence every day.

A physician orders tests before making a serious diagnosis. An engineer calculates whether a bridge is safe. A judge asks for proof. A scientist expects a claim to be examined and corrected.

Yet the same person may accept a miracle story without reliable evidence, consult astrology before a wedding, fear an unlucky date, or tell a child not to question religion.

Evidence governs one part of life.

Inherited belief governs another.

This divided mind is one of the great contradictions of the modern world.

We live in an age of vaccines, genetic science, satellite communication, and space exploration. Yet ancient fears remain active inside modern minds.

A person may trust physics while flying across the world and consult an astrologer after landing. He may understand biology in the classroom but defend a sacred creation story at home. He may rely on modern medicine during illness and still believe that ritual or divine favor determined the outcome.

This book asks why.

Why do intelligent people accept claims from their own religion that they would reject if the same claims came from another faith?

Why are religious conclusions taught before children are old enough to examine them?

Why does doubt, the beginning of knowledge, become a moral danger in the house of faith?

Religion: The Enemy of Reason examines how religion feeds superstition, weakens science, and silences questioning. It explores how belief enters the mind early, attaches itself to family and identity, gains power through fear, and receives protection from the standards used everywhere else.

Science has not answered every question. It may never answer all of them. Mystery remains. Wonder remains. Human knowledge remains incomplete.

Wonder is not the enemy of reason.

The danger begins when someone commands us to stop asking.

Why This Question Became Personal

I grew up in a Hindu religious family.

Religion first entered my life through people I loved. It came through family rituals, festivals, recited verses, lamps, sacred images, and the seriousness on the faces of adults.

Long before I could evaluate religious claims, I understood that they were not ordinary ideas. They were to be respected. Questioning them felt different from questioning anything else.

My journey away from inherited belief was gradual.

I did not wake one morning and reject everything I had been taught. The process was slower and more difficult. Questioning religion did not feel like correcting an abstract mistake. At times, it felt like stepping away from memories and people I still valued.

I had to learn that love for those who taught me religion did not require acceptance of everything they believed.

My work as a physician strengthened that lesson.

During more than thirty-five years in medicine, I have repeatedly seen that reality does not change because a belief is sincere or comforting.

A wrong diagnosis does not become correct because it is traditional. A treatment does not become effective because a respected elder recommends it. A claim does not become true because millions repeat it.

The body responds to biology, not conviction.

In medicine, we listen, examine, test, treat, observe, and correct. When better evidence appears, a responsible physician changes the diagnosis or treatment. Refusing to change because an earlier belief feels familiar can harm the patient.

Medicine also teaches humility.

A physician does not always know the answer. Treatments fail. Diagnoses remain uncertain. New evidence replaces old understanding. The honest words "I do not know" are often safer than false certainty.

The same principle applies beyond medicine.

False beliefs can wound a child's mind, restrict a woman's freedom, preserve caste humiliation, divide communities, and teach people to obey authority against their conscience.

I do not ask readers to accept my conclusions because I am a physician or because my name appears on the cover.

Question what I write.

Compare it with evidence.

Notice where I may be incomplete, unfair, or wrong.

No author should ask for the intellectual obedience this series criticizes in religion.

A Child Inherits Religion Before He Can Choose

No child is born knowing the name of a god.

A newborn does not arrive with a scripture, a prayer, a caste, a food restriction, a purity rule, or a fear of hell. The child does not know which temple, mosque, church, synagogue, monastery, or shrine the family will visit.

All of that comes later.

It comes through parents, grandparents, clergy, teachers, rituals, songs, festivals, stories, and sacred books. It comes so early that religion becomes difficult to separate from love, safety, family memory, and belonging.

Most people do not choose their first religion after comparing evidence.

They inherit it.

The same child would probably grow up believing something very different if born into another family, country, or century.

A Hindu child may learn that certain gods and rituals are sacred. A Muslim child may receive a different understanding of God and revelation. A Christian child may be taught that salvation depends on Jesus. Jewish, Sikh, Buddhist, Jain, and other children receive still other maps of reality.

Each community may speak with confidence.

Each may say, “This is the truth.”

This should make every honest person pause.

When belief depends so strongly on birthplace, family, language, and childhood instruction, how can we know that it came from a divine source rather than human inheritance?

This question does not prove that every religious claim is false.

It does show that familiarity is not evidence.

A belief does not become universal truth because it was taught first, repeated often, or connected to people we love.

A child belongs to humanity before he belongs to any religion.

He should be allowed to learn how to think before adults decide what he must believe.

This Is Not an Attack on Believers

This series is not written to mock ordinary religious people.

It is not an attack on the grieving person who prays after a death, the frightened patient who seeks hope, or the lonely person who finds community in a place of worship.

Illness, death, loneliness, poverty, uncertainty, and grief are real. People naturally reach for whatever helps them endure.

A humane criticism of religion must understand this.

Many religious people are kind, generous, honest, and deeply moral. They care for families, help neighbors, feed the hungry, comfort the dying, and serve their communities.

Religion has also inspired art, music, architecture, poetry, charity, festivals, and social belonging.

These contributions should be acknowledged.

But goodness does not prove theology.

Kind people can believe false things. A beautiful tradition can grow around a mistaken explanation. A ritual can carry emotional meaning without possessing supernatural power. A sacred book can contain wisdom without having been written by a god.

The purpose of this series is not to judge the personal worth of believers.

It is to examine religion as a system of claims and authority.

Religion teaches children. It influences marriage, medicine, gender, morality, law, education, and politics. It tells people what God wants, which books are sacred, which behaviors are sinful, and what may happen after death.

Any system with that much power must remain open to examination.

Respect for people should never require silence about ideas.

From Shelter to Authority

Religion did not arise without human need.

Early human beings lived in a dangerous and poorly understood world. Lightning split the sky. Floods destroyed settlements. Disease took children. Childbirth endangered mothers. Death arrived without explanation or effective treatment.

Human beings created stories to make this world less frightening.

Gods were placed behind storms. Spirits were placed behind disease. Demons explained disturbed behavior. Ritual offered a sense of action when practical control was limited. Prayer gave people something to do when no medicine was available.

Religion became an early shelter of meaning.

It helped communities endure uncertainty. It offered language for grief and death. It created social identity, shared ceremony, moral rules, music, art, and architecture.

We should acknowledge this history honestly.

But usefulness and truth are different.

A story may calm fear without accurately describing reality. A ritual may provide emotional relief without changing the event. A belief may help someone survive a painful night and still be false.

Humanity now possesses better tools for understanding the world: science, medicine, psychology, history, secular law, and modern ideas of human rights.

We no longer need divine anger to explain epidemics. We no longer need possession to explain schizophrenia or epilepsy. We no longer need supernatural punishment to explain infertility, disability, or miscarriage.

The problem is not that religion once provided explanations.

The problem begins when those explanations are made sacred and protected from correction.

An old shelter becomes a prison when people are forbidden to walk outside it.

Truth Should Not Change at the Border

Religions do not merely disagree about minor rituals.

They make conflicting claims about God, prophets, scriptures, creation, salvation, rebirth, heaven, hell, purity, food, gender, sexuality, marriage, and death.

A practice considered sacred in one religion may be forbidden in another. One tradition teaches a single life followed by judgment. Another teaches repeated rebirth. Each may present its answer as final.

Billions of people reject every one of these supposedly final answers.

Scientific knowledge behaves differently.

Gravity does not become Hindu in India, Christian in America, or Muslim in Saudi Arabia. Germs do not follow scripture. DNA does not change according to prayer. A vaccine does not work only for one caste or religion.

Science is not infallible. Scientists can be biased, dishonest, ambitious, and wrong.

Its strength lies in its method.

Scientific claims must answer to reality. They can be tested by people from different nations, cultures, and religions. When better evidence appears, conclusions can change.

Religion usually begins with a conclusion that must be protected.

Science exposes its conclusions to correction.

A truth that changes whenever a child is born across a border looks less like universal revelation and more like inherited human tradition.

When Belief and Human Dignity Collide

The moral foundation of this series is simple:

When a belief and a human being come into conflict, protect the human being.

If a child is terrified by hell, the child matters more than the doctrine.

If a woman is treated as impure, her dignity matters more than the ritual.

If caste humiliates a person because of birth, equality matters more than tradition.

If a patient is blamed through sin, karma, or weak faith, the patient matters more than the explanation.

If a person is threatened for disbelief, freedom of conscience matters more than the anger of the majority.

Religious language can make harmful practices sound honorable.

Exclusion becomes purity.

Submission becomes duty.

Fear becomes faith.

Silence becomes respect.

We must look behind those words and ask what the practice does to actual people.

Does it reduce suffering or increase it?

Does it encourage inquiry or silence it?

Does it protect freedom or demand obedience?

No scripture, ritual, institution, or god should be protected more carefully than the person harmed in its name.

The human being must come first.

Ten Books, One Investigation

Religion affects humanity through many connected systems. That is why **Religion: The Holy Poison** is a ten-book series:

1. **Religion: The Enemy of Reason**
2. **Religion: The Mass Delusion**
3. **Religion: The Child's First Prison**
4. **Religion: The Biggest Addiction**
5. **Religion: The Business of Faith and Fear**
6. **Religion: The Obedience Machine**
7. **Religion: The Poisoning of Politics**
8. **Religion: The Weapon of Hatred**
9. **Religion: The Justification of Inequality**
10. **Religion: The Curse on Humanity**

Each volume examines one part of the same system: reason, inherited belief, childhood conditioning, emotional dependence, religious wealth, obedience, political power, hatred, inequality, and the accumulated human cost.

Together, these books ask one central question:

What has religion done to humanity?

Not only what it has promised.

Not only what beauty it has produced.

Not only how many people believe.

What has it done to the human mind, the child, the family, the body, the outsider, and the future?

That question deserves an honest answer.

Read With Courage

Some ideas in this series may feel uncomfortable.

They may touch childhood memories, family loyalty, cultural identity, grief, hope, and fears planted before reason had a voice.

That discomfort deserves patience.

It should not end the examination.

When uneasiness appears, ask where it comes from. Is the argument factually weak, or has questioning itself been made frightening? Is the objection based on evidence, or on fear of what disbelief might mean?

Do not accept an argument merely because it supports your present view.

Do not reject it merely because it disturbs you.

Most people can identify the contradictions and superstitions of other religions more easily than those of their own. Intellectual honesty begins when we apply the same standard at home.

If a religious claim is true, it should survive examination.

If a tradition is humane, its value should remain after fear is removed.

If a doctrine cannot withstand evidence, moral scrutiny, and honest questioning, humanity should not remain obedient to it merely because it is old or sacred.

Question this book as carefully as it questions religion.

Protect the person before the belief.

Do not believe me.

Follow the evidence.

Chapter 1: The First Forbidden Question

A child asks, “Who made God?”

The adults answer, “No one. God has always existed.”

The child thinks for a moment.

“If God does not need a creator, why does the universe need one?”

The room changes.

Someone laughs and changes the subject. Another adult says that God is beyond human understanding. A parent may warn the child not to be disrespectful.

The response may be gentle. No one needs to shout. The child still understands that he has crossed a line.

His first question was tolerated.

His second question was not.

He has discovered that religious answers are treated differently from ordinary answers.

In school, a question may reveal confusion or improve understanding. In medicine, a question can prevent an error. In science, a question may begin a discovery.

But in religion, the same question can be treated as a threat.

The child is not attacking anyone. He is trying to understand. Yet without realizing it, he has challenged one of religion's most important protections:

the right of sacred claims to escape ordinary examination.

That is often the first forbidden question.

Curiosity Comes Before Belief

Children begin life as questioners.

They ask why the sky changes color, where babies come from, why people grow old, why animals suffer, and what happens after death. They notice contradictions adults have learned not to see.

This curiosity is not rebellion.

It is how the mind grows.

A healthy answer does not require an adult to know everything. "I do not know" may be more honest than a confident invention. A parent can say, "Different people believe different things," or "Let us think about that together."

Such answers teach the child that uncertainty is safe.

Religious instruction often takes a different path.

The child is given conclusions before he has the ability to evaluate them:

God exists.

God sees everything.

Prayer works.

The sacred book is true.

Good people are rewarded.

Doubters may be punished.

These ideas arrive through people the child trusts. They are wrapped in family affection, ritual, music, celebration, and moral instruction. By the time the child is old enough to examine them, they may already feel like part of reality.

The issue is not that parents deliberately deceive their children. Most pass forward what they sincerely believe. They think they are giving the child morality, protection, identity, and belonging.

But the order of learning has been reversed.

Normally, we examine a claim and then decide whether it deserves confidence.

Religion often asks for confidence first and permits examination later, if examination is permitted at all.

When Love Teaches Silence

Religion rarely silences children through argument.

It makes questioning emotionally costly.

A child depends on the approval of parents and grandparents. He wants to be loved, praised, and included. If religious belief is tied to family loyalty, questioning the belief can feel like rejecting the family.

The lesson may be conveyed through small reactions.

A smile disappears.

An elder becomes disappointed.

A parent says, "Why are you talking like this?"

A teacher warns that too much questioning can weaken faith.

No formal punishment is necessary.

The child learns what brings warmth and what creates tension.

This is how silence can begin to look like goodness.

The obedient child repeats the prayer, follows the ritual, respects the sacred book, and avoids embarrassing questions. Adults may describe him as well raised, devoted, or moral.

The questioning child may be called stubborn, arrogant, confused, badly influenced, or disrespectful.

In some homes, the warning is stronger. Doubt may anger God. It may bring punishment, weaken the family, or endanger the child's soul.

The child soon understands that belief is rewarded not only as truth, but as virtue.

This matters because children do not separate ideas from relationships as easily as adults imagine. When a belief comes through love, rejecting the belief may feel like rejecting the person who taught it.

That pressure can survive for decades.

An adult may continue attending ceremonies he no longer accepts. He may repeat words he does not believe because honesty would upset a parent. His obedience appears voluntary from outside, yet it is shaped by guilt, fear, and affection.

Respect for people matters.

But respect should never require the surrender of judgment.

Faith as an Answer

When a religious explanation reaches its limit, the word *faith* often appears.

"How do we know God exists?"

"You must have faith."

"How do we know this scripture came from God?"

"Faith does not require proof."

"How do we know prayer changes events?"

"You must believe."

To a child, faith may sound like a deeper form of knowledge.

Later, he may recognize that faith often means accepting a claim without enough evidence.

The weakness becomes clearer when religions are compared.

A Hindu child may grow up believing that particular gods and rituals are sacred. A Christian child may be taught that salvation comes through Jesus. A Muslim child may learn that the Quran is the final revelation. A Sikh child may revere the Guru Granth Sahib.

Each child may be surrounded by loving adults who speak with complete confidence.

Their confidence cannot make all their conflicting claims true at the same time.

Faith may explain why a person trusts a belief.

It does not establish that the belief corresponds to reality.

This leads to one of the most useful questions a believer can ask:

Would I believe this if I had been born into another religion?

The question does not settle every theological issue.

It reveals how much certainty may have come from inheritance rather than independent examination.

A child does not choose the religious world into which he is born. Yet he may later be told that accepting it is virtue and doubting it is failure.

That is not a fair beginning.

The Unequal Rules of Belief

Children notice when adults use different standards.

A parent may warn against rumors. A teacher may ask for evidence. A doctor may order tests. A court may require proof before punishment.

In ordinary life, caution is considered wise.

Then religion makes claims of enormous importance:

God exists.

A soul survives death.

Prayer changes events.

Miracles occur.

A sacred book contains divine truth.

Certain acts bring punishment after death.

A ritual protects the family.

When the child asks how these things are known, the standards change. Evidence is replaced by faith, tradition, authority, or personal feeling.

The child may not know the phrase *double standard*, but he can sense one.

Adults who are skeptical about another religion's miracle may accept their own. People who question another scripture may protect the one they inherited. A family that rejects one community's superstition may obey a similar belief at home.

The same evidence is judged differently depending on whose religion it supports.

If this division remains unexamined, it can shape the adult mind.

The person may become careful and evidence-driven in professional life while preserving a protected religious area where ordinary scrutiny does not enter.

The contradiction begins early:

Evidence has rules.

Religion has exceptions.

Doubt Is Not a Moral Failure

Religion often gives doubt a bad reputation.

Doubt may be described as weakness, pride, temptation, disloyalty, or sin. A believer may feel guilty before he has even examined the question that disturbed him.

The emotional punishment arrives before the reasoning begins.

This is highly effective.

Once people fear doubt itself, a belief can survive without answering criticism.

Yet doubt is essential wherever truth matters.

During medical practice, I learned that a question asked at the right moment could change a diagnosis. A physician who becomes certain too quickly may overlook the evidence that does not fit. Doubt protects the patient by keeping the conclusion open.

The same principle applies in other fields.

A judge hesitates before convicting.

A scientist tries to disprove a hypothesis.

A parent checks a rumor before punishing a child.

Doubt does not mean rejecting everything.

It means allowing the possibility of error.

That possibility protects us.

Without doubt, confidence can become dangerous. Tradition appears infallible. Authority becomes difficult to challenge. A person may continue believing not because the claim is strong, but because he has been taught that questioning it is wrong.

The child who doubts is not losing morality.

He is developing judgment.

If a belief is true, honest examination should not destroy it.

If it is false, doubt may be the only path out.

What Happens When a Child Learns to Stop

The first surrender of a question may seem harmless.

The child stops asking who made God. He repeats the prayer. Family life continues.

But habits of thought rarely stay confined to one claim.

Once a person learns that sacred ideas do not require evidence, the protected area can expand. Scripture, miracles, astrology, ritual danger, purity rules, divine punishment, possession, and karma may all enter under the same shelter.

The child does not automatically accept every superstition.

He has learned the method that allows superstition to survive:

some claims are too sacred, comforting, or frightening to examine normally.

The consequences may appear much later.

A family consults astrology before marriage.

A patient seeks ritual help before medical care.

A girl accepts exclusion during menstruation because everyone calls it tradition.

An adult remains frightened of hell even after the doctrine no longer makes sense to him.

These later harms do not arise from one childhood conversation alone. Human belief is more complicated than that.

But the early lesson matters.

When sacred authority speaks, the mind should step back.

Over time, a boundary becomes a habit, and a habit becomes part of identity.

The Right to Arrive

The goal should not be to replace religious indoctrination with atheist indoctrination.

A child should not be ordered to disbelieve any more than he should be ordered to believe.

He should be taught how to think.

He should learn that different religions make conflicting claims. He should understand that confidence can be inherited and that respected adults can be mistaken. He should be allowed to ask what evidence supports a belief and what evidence might weaken it.

Most importantly, he should know that love does not depend on agreement.

A parent can say:

“You are free to question.”

“I may not know the answer.”

“You do not have to pretend.”

“We can disagree and still love one another.”

These sentences give a child something religion too often withholds:

the freedom to arrive at his own conclusion.

That freedom does not guarantee that he will become an atheist. He may remain religious, revise his belief, or remain uncertain.

The outcome belongs to him.

A belief reached freely has more dignity than one preserved through fear.

The forbidden question often returns later.

It may return in a science class, during a medical crisis, after reading another religion’s scripture, or when a person watches a harmful custom imposed on someone vulnerable.

It may return when his own child asks the question he once learned to suppress.

At that moment, he can silence it again.

Or he can finally listen.

Intellectual freedom begins when the question is allowed to stand without fear.

The child asks, "Who made God?"

Behind that question lies another:

How do we know any religious claim is true?

This book follows that question wherever it leads.

Key Points

- Children question naturally; doubt is not rebellion.
- Religious beliefs are often taught before children can evaluate them.
- Family affection and approval can make questioning emotionally costly.
- Faith may explain trust, but it cannot settle conflicting religious claims.
- Religion often receives weaker standards of evidence than ordinary claims.
- Doubt protects the mind by allowing the possibility of error.
- Children should be given the freedom and tools to reach their own conclusions.

Reader Reflection

What religious question was I discouraged from asking?

Would I hold the same beliefs if I had been born into another family or religion?

Which inherited belief have I never examined by the standards I use elsewhere?

Transition to Chapter 2

Religion can discourage questioning from the outside, but belief also grows from tendencies already present within the human mind.

We search for patterns, imagine intention behind events, remember confirming stories, and become less skeptical when we are afraid.

To understand why religious explanations feel so natural, we must first understand the brain that receives them.

Chapter 2: The Brain That Finds Gods

You are walking alone after dark.

A sound comes from the bushes.

Your body reacts before your mind has identified the cause. You stop. Your muscles tighten. You turn toward the movement, ready to run or defend yourself.

Perhaps an animal is hiding there.

Perhaps another person is watching.

Perhaps it was only the wind.

For most of human history, hesitation could be costly. An ancestor who mistook the wind for a predator wasted a few moments. One who mistook a predator for the wind might not survive.

The human brain therefore became highly sensitive to possible danger and hidden agency. It often assumes that something or someone caused an unexplained event before complete evidence is available.

That instinct helped keep human beings alive.

It also made invisible agents easy to imagine.

A storm could seem angry. Disease could feel like punishment. A sudden recovery could appear to be intervention. A dream could become a warning. A coincidence could look like a message.

Behind the event, the mind places an intention:

a spirit,

an ancestor,

a demon,

a god,

a curse,

or a hidden plan.

Religion fits naturally into these habits of thought.

This does not mean religious people are foolish. The same brain that discovers mathematics, composes music, performs surgery, and sends machines into space can also find intention where none has been demonstrated.

The problem is not a defective mind.

It is a human mind doing what it evolved to do, sometimes in situations for which its ancient instincts are poorly suited.

A Brain Built for Survival

The brain did not evolve primarily to discover perfect truth.

It evolved to help an organism survive, reproduce, cooperate, and respond quickly to danger.

Accuracy matters, but speed often matters too.

A person standing in a forest cannot conduct a controlled experiment before reacting to a sudden movement. The brain uses shortcuts. It makes rapid judgments from incomplete information and fills gaps with what seems most likely.

These shortcuts remain active in modern life.

A patient develops chest pain, and a physician immediately considers several dangerous causes. This rapid pattern recognition can save a life. But medical training also teaches that the first impression may be wrong. Similar symptoms can arise from very different illnesses.

The pattern must be checked.

Tests, examination, observation, and response to treatment help distinguish a correct diagnosis from a convincing mistake.

I have seen how easily an early impression can become fixed. Once a diagnosis feels right, contradictory evidence may be overlooked. The responsible physician must therefore ask not only, "What supports my conclusion?" but also, "What does not fit?"

The same discipline is needed in religious belief.

A pattern may feel real before it has been demonstrated.

A cause may feel obvious because it satisfies expectation.

A story may feel complete because the alternative is uncertainty.

The mind reaches an answer.

Reason must then ask whether the answer is true.

The Pattern That May Not Be There

Pattern recognition is one of the brain's greatest strengths.

We learn that dark clouds may bring rain, that certain symptoms may indicate disease, and that particular actions often lead to predictable consequences. Without patterns, experience would remain a collection of unrelated events.

But the mind sometimes connects events that happened close together without establishing that one caused the other.

A person prays, and good news arrives the next day.

A family performs a ritual, and a child's fever improves.

A student wears a special ring during an examination and passes.

Someone dreams of a relative and receives a telephone call from that person later.

The sequence feels meaningful.

Yet one event following another does not prove causation.

The fever may already have been improving. The student may have prepared well. The telephone call may have been coincidence. Among thousands of dreams and daily events, some are bound to resemble what happens later.

Coincidence becomes especially persuasive when it touches hope or fear.

An ordinary probability becomes a personal story.

Religion supplies the interpretation:

The prayer was answered.

The ritual worked.

The ring brought luck.

The dream was a sign.

The interpretation feels richer than chance. “God sent a message” carries emotional weight. “These events happened near each other” sounds empty by comparison.

But emotional richness does not establish a causal connection.

The mind must learn to distinguish between a pattern discovered in reality and a pattern created by expectation.

Someone Must Have Done It

Human beings naturally search for agents.

When a door opens, we wonder who opened it. When a child cries, a parent looks for a cause. When someone acts strangely, we try to understand what he intended.

This ability is essential in social life.

It becomes less reliable when extended to events that may have no conscious agent behind them.

A crop fails, and someone asks which deity was offended.

A child has a seizure, and the family suspects possession.

A storm destroys a village, and a preacher describes divine anger.

Several misfortunes occur close together, and relatives begin speaking of a curse.

These explanations place a mind behind the event.

For many people, even a hostile supernatural agent feels easier to face than impersonal nature. If a god, spirit, or curse caused the problem, perhaps the cause can be negotiated with.

A ritual can be performed.

An offering can be made.

A specialist can be consulted.

Something can be done.

Natural explanations are often less emotionally satisfying. Infection, genetics, weather, brain disease, human error, and chance may involve complicated chains of causes. They do not promise that anyone is listening or that suffering has a purpose.

The desire for an agent is understandable.

It is not evidence that an agent exists.

A storm can destroy without anger.

A disease can strike without punishment.

A child can suffer without being cursed.

Nature does not need intention in order to produce tragedy.

The Mind Counts the Hits

Once a person accepts a belief, the mind becomes more likely to notice events that support it.

A prayer appears to be answered, and the story is remembered for years.

Many unanswered prayers disappear from memory.

An astrologer gives several vague predictions. The one that seems accurate is repeated at family gatherings. The others are forgotten.

A ritual is followed by good news, and the ritual receives credit. When nothing happens, the failure attracts little attention.

This is not usually deliberate dishonesty.

Memory is selective.

Striking events remain vivid. Ordinary failures do not. Human beings tell stories about surprising success, not about the many occasions when nothing unusual happened.

Religious communities strengthen this tendency by repeating accounts of healing, protection, answered prayer, prophecy, and divine signs.

The unsuccessful cases rarely become part of the tradition.

Over time, selected memory creates the impression of overwhelming evidence.

Honest evaluation requires a different habit.

Count the misses as well as the hits.

If prayer changes outcomes, all prayers matter.

If astrology predicts, all predictions matter.

If a ritual protects, the people who performed it must be compared with those who did not.

The complete record may look very different from the remembered story.

This is why reliable knowledge depends on documentation, comparison, and testing. Memory alone is too easily persuaded by what the mind hopes to find.

Fear Lowers the Standard

A person who dismisses superstition in ordinary circumstances may become cautious when someone he loves is in danger.

A father may laugh at the evil eye until his baby develops a fever.

A young woman may reject astrology until an astrologer predicts danger to her marriage.

A family may doubt spiritual healing until a physician says that the illness is difficult to treat.

Then someone says:

"I do not fully believe it, but why take a chance?"

No new evidence has appeared.

The emotional cost of refusing has increased.

Fear makes weak claims feel more serious because the imagined consequence of disbelief becomes frightening.

Religion gains much of its power during illness, grief, childbirth, financial insecurity, marriage, old age, and death. At such moments, people are not calmly comparing explanations. They are searching for safety.

A ritual provides action.

A promise provides reassurance.

A religious authority provides direction.

The resulting relief may be genuine. Prayer can calm a frightened person. A familiar ceremony can give a family something to do when it feels helpless.

But reduced anxiety does not prove that the invisible danger was real.

A charm may calm a parent while the illness continues.

A prayer may strengthen a patient emotionally while medicine does the healing.

The relief is real.

The supernatural explanation remains unproven.

Why Familiar Beliefs Feel True

Religious ideas rarely enter the mind as isolated propositions.

They arrive through trusted people, familiar language, music, stories, celebrations, and family life. A belief repeated from early childhood can begin to feel like an obvious feature of the world.

Other religions feel different because they are unfamiliar.

A person may immediately recognize the human origins of another tradition's god, miracle, or ritual. His own tradition feels natural because it is tied to memory and belonging.

Familiarity is psychologically powerful.

It is not the same as evidence.

A belief may feel self-evident simply because the mind has heard it thousands of times.

This is why birthplace matters so much in religion. The same human brain, raised in another culture, may experience a very different theology as equally natural and unquestionably true.

The feeling of certainty is therefore not a reliable guide to which religion, if any, accurately describes reality.

It may reveal what the mind learned first.

The Mind Protects What It Loves

Human intelligence does not always correct bias.

Sometimes it gives bias better arguments.

A thoughtful person can build elaborate defenses for a belief learned in childhood. He can reinterpret failures, explain contradictions, and develop sophisticated philosophical language for conclusions that were emotionally established long before the reasoning began.

The argument may be new.

The belief is old.

This is rationalization: using reasoning to protect a conclusion rather than to discover whether the conclusion is true.

Religion is especially vulnerable to this because belief is connected to identity, family, death, hope, and morality. Examining it may feel like threatening much more than an idea.

We recognize this process easily in another religion.

We see how another believer explains away a failed prophecy, protects a miracle, or treats his scripture differently from every other book.

It is harder to see the same process in ourselves.

The important question is not merely:

"Why do people believe strange things?"

It is:

Which of my own beliefs have I protected because I want them to be true?

That question turns psychology into self-examination.

Explanation Is Not Refutation

There is an important caution.

Explaining why people are psychologically inclined toward religious belief does not, by itself, prove that no god exists.

A belief can have a natural psychological origin and still happen to be true. Human beings are psychologically inclined to believe that other minds exist, and other minds do exist.

It would therefore be a mistake to argue:

"Religion arises naturally from the brain, so every religious claim must be false."

That conclusion does not follow.

Psychology explains why a belief may feel compelling. It does not settle the external question.

Whether a god exists must still be evaluated through the evidence offered for that claim.

The psychological explanation matters because it removes one mistaken argument: the widespread and persistent nature of religion does not itself demonstrate that religion is true.

A belief may spread because it fits human fear, memory, pattern recognition, and social life.

The fact that many minds independently produce similar supernatural explanations tells us something important about minds.

It does not automatically tell us that the supernatural beings exist.

The origin of a belief and the truth of a belief are separate questions.

Both must be examined carefully.

Psychological Need Is Not Evidence

Religion fits many human needs.

It offers intention behind the universe, meaning behind suffering, action against helplessness, community during loneliness, and reassurance in the face of death.

This helps explain why religion appears across cultures.

But the presence of a need does not demonstrate that the belief satisfying it is true.

The desire for immortality does not establish an afterlife.

The fear of meaninglessness does not prove a cosmic purpose.

The comfort of believing that someone watches over us does not demonstrate that anyone does.

A hungry person's need for food points toward something that can be independently shown to exist. The emotional need for eternal life has no similar evidentiary force.

The feeling may be profound.

The conclusion still requires support.

Religion becomes intellectually dangerous when psychological need is allowed to answer a factual question.

"I need this to be true" becomes "therefore it is true."

A careful mind keeps those statements separate.

Learning to Correct the Mind

The human brain is vulnerable, but it is not helpless.

The same mind that invents patterns can learn probability.

The same mind that imagines invisible agents can investigate natural causes.

The same memory that preserves confirming stories can keep complete records.

The same person who wants certainty can learn to wait.

Reason is not the absence of mental weakness.

It is the discipline of correcting for it.

We can pause before assigning meaning to coincidence.

We can ask what evidence would change our minds.

We can compare our own religion with others using the same standard.

We can notice when fear has lowered the quality of our judgment.

This takes practice.

A belief learned early may continue producing anxiety even after it has been rejected intellectually. A person may know that astrology lacks evidence and still feel uneasy after hearing an unfavorable prediction. Someone may stop believing in hell yet carry the fear of it for years.

The persistence of emotion does not prove the belief.

It shows how deeply the mind was trained.

Freedom begins when we stop treating every powerful feeling as information about reality.

Fear tells us that something matters to us.

It does not tell us that the feared claim is true.

The brain that finds gods can also examine why it found them.

That capacity for self-correction is one of the strongest foundations of intellectual freedom.

Key Points

- The human brain evolved for survival, not perfect accuracy.
- Pattern recognition can reveal real connections, but it can also turn coincidence into apparent meaning.

- Human beings naturally imagine agents and intentions behind events that may have natural causes.
- Memory favors striking confirmations and often neglects failures.
- Fear lowers the standard of evidence and makes rituals or supernatural explanations more attractive.
- Familiarity can make an inherited belief feel self-evidently true.
- Explaining the psychology of belief does not by itself disprove God, but it does show why widespread belief is not sufficient evidence.
- Reason helps the mind correct for its own natural weaknesses.

Reader Reflection

Which belief feels true mainly because it has always been familiar?

Do I remember apparent answers to prayer more clearly than unanswered prayers?

What belief might I defend because I want it to be true?

Transition to Chapter 3

The human mind is prepared to find patterns, hidden agents, and meaningful explanations. Religion becomes even more powerful when it satisfies another deep need: relief from uncertainty.

Where life gives incomplete answers, religion offers certainty.

The next chapter examines why certainty feels so comforting, and why that comfort can become dangerous.

Chapter 3: The Sweet Poison of Certainty

A child dies.

Relatives gather in the family home. Voices become quiet. Food is prepared but barely touched. The parents sit among people who love them, yet no one can enter the depth of their loss.

Someone finally speaks.

"God needed another angel."

Another person says, "Everything happens for a reason."

Elsewhere, the words may be different.

It was karma.

It was written.

It was God's will.

It was a test.

The child has gone to a better place.

These sentences are usually offered with kindness. People feel helpless before grief, and silence can seem unbearable. They want to give the parents something to hold.

Sometimes the words comfort.

Sometimes they wound.

A grieving mother may not be ready to hear that her child's death served a higher purpose. A father may find no comfort in being told that someone planned the tragedy. Yet religious certainty often arrives before anyone asks whether an explanation is needed.

I understand the temptation.

When suffering has no clear meaning, a confident answer can feel like shelter. Human beings want to believe that someone is in control, that injustice will eventually be corrected, and that death is not final.

Religion offers this reassurance with remarkable confidence.

It explains why we exist, why we suffer, what happens after death, which rules are eternal, and how the universe is governed.

Where life remains silent, religion speaks.

The answer may feel sweet because uncertainty is painful.

That sweetness is what makes false certainty so powerful.

Why We Want Final Answers

Uncertainty is not merely an intellectual discomfort.

We feel it in the body.

A patient waits for biopsy results. A parent watches a sick child. A young couple wonders whether a pregnancy will continue safely. An older person thinks about death.

The mind searches for a definite answer because not knowing creates tension.

A clear explanation reduces anxiety, even when the explanation is poorly supported.

This is why people consult astrologers before marriage, seek miracle workers during illness, interpret misfortune through karma, and cling to promises about the afterlife.

The belief gives uncertainty a shape.

It says:

There is a plan.

Someone is in control.

The suffering has meaning.

The future can be known.

This reassurance is emotionally attractive.

Religious rituals can also provide structure. During grief, they give families familiar words and actions. During illness, they create a sense of participation. During death, they place the loss inside a larger story.

These benefits may be genuine.

But emotional usefulness and factual truth are different questions.

A belief can help someone endure a painful night without accurately describing the universe.

The danger begins when “I need this to be true” quietly becomes “therefore it is true.”

The Answer That Arrives Too Quickly

Certainty can become cruel when it explains another person’s suffering before listening to it.

A woman loses a pregnancy and is told that God had another plan.

A man develops cancer and hears that he is being tested.

A poor family is advised to accept hardship as karma.

A person with depression is told that stronger faith would bring peace.

These explanations may protect the belief system more than the person.

If a just God controls everything, then suffering must somehow be just.

If karma governs life, pain must have been earned.

If prayer works, unanswered prayer must have a hidden purpose.

The doctrine remains safe.

The sufferer carries the burden of making the tragedy reasonable.

During medical practice, I have often spoken with families when certainty was impossible. A diagnosis was unclear. Treatment had failed. The future could not be predicted honestly.

In those moments, false reassurancc would not havc bccn kindness.

Sometimes the most compassionate words were simple:

“I am sorry.”

“This is terribly difficult.”

“I do not know why this happened.”

“I am here.”

There is dignity in not explaining another person’s pain.

Compassion asks what the person needs.

Certainty too often asks how the doctrine can be preserved.

When Certainty Stops Inquiry

A final answer does more than comfort.

It can close the investigation.

If illness is called punishment, attention may move away from biology and treatment.

If poverty is described as fate, social injustice may be ignored.

If scripture has already settled a moral question, the harm caused by the rule may no longer be examined.

Certainty replaces inquiry with conclusion.

Consider the phrase, “Everything happens for a reason.”

Sometimes people use it simply to express endurance. At other times, it suggests that every tragedy was intentionally arranged.

If every event belongs to a perfect plan, human responsibility can become blurred.

Why investigate a preventable death?

Why challenge an unjust custom?

Why ask whether care was delayed?

Why reform a system if suffering has already been given cosmic meaning?

Religious explanations often direct attention upward, toward God, fate, karma, or divine purpose.

Reason brings attention back to causes we can examine.

A child’s death from infection may reveal the need for vaccination, clean water, or earlier treatment.

A woman’s suffering may expose violence or inequality.

A poor person’s hardship may reflect present injustice, not past-life karma.

Natural explanations may feel less comforting because they do not promise that everything belongs to a plan.

But they may reveal what can be prevented.

An answer that reduces responsibility should always be treated with caution.

Certainty Becomes Authority

Certainty is especially dangerous when people believe they know what God wants from others.

A believer may describe himself as humble before God while speaking confidently about who is saved, which behavior is sinful, which scripture is final, and which questions are forbidden.

Private faith becomes authority over another person's life.

This often appears in matters of gender, sexuality, caste, family obedience, and unbelief.

A parent frightens a child because he is certain hell is real.

A family restricts a daughter because it is certain purity matters.

A community silences a doubter because it is certain disbelief threatens moral order.

The people involved may not consider themselves cruel. They may believe they are protecting souls, preserving tradition, or obeying divine law.

That is precisely the danger.

Ordinary moral hesitation can disappear when an action carries sacred approval. A person who might otherwise ask, "Am I causing harm?" continues because he believes the answer has already been given by God.

Certainty can make compassionate people behave without compassion.

This is not unique to religion. Political ideologies and secular movements can also become rigid and cruel when treated as infallible.

Religion adds another layer because its authority is placed beyond ordinary human correction.

A political claim may be criticized as human.

A sacred command may be defended as the will of the universe.

The more certain we feel, the more important it becomes to ask:

What if I am wrong, and someone else must pay the price?

That question restores moral caution.

The Honesty of "I Do Not Know"

Religion sometimes presents uncertainty as a weakness of science.

Scientific conclusions change. Experts disagree. New evidence replaces older understanding.

To someone seeking certainty, this can look unreliable.

In reality, the willingness to change is one of science's greatest strengths.

A physician revises a diagnosis when the evidence no longer fits.

A scientist modifies a theory when new findings demand it.

A historian changes an interpretation when new records appear.

Correction is not failure.

It is the method working.

Religion often offers greater confidence because its central answers are protected. Yet an answer is not more reliable merely because it cannot be revised.

Permanence can preserve error as easily as truth.

The words "I do not know" require courage because they leave us temporarily exposed. They do not promise an afterlife, explain every tragedy, or guarantee that justice will prevail.

But they protect us from invention.

"I do not know" does not end inquiry.

It begins it.

If I do not know why a child died, I will not tell the parents that God wanted it.

If I do not know what happens after death, I will not frighten children with confident descriptions of hell.

If I do not know the purpose of the universe, I will not claim that another person's suffering, caste, gender, or unbelief was divinely assigned.

Honest uncertainty limits authority.

That is one reason systems built on certainty resist it.

What Certainty Hides

Certainty often conceals the weakness of the evidence beneath the strength of the emotion.

A person may say, "I know God answered my prayer."

What he may truly know is that he prayed and something happened afterward.

He may say, “I know this scripture is divine.”

What he may truly know is that he was taught to revere it.

He may say, “I know the ritual protects us.”

What he may truly know is that refusing the ritual creates fear.

Confidence can feel like knowledge while resting on memory, culture, expectation, or need.

The distinction matters because confidence persuades others.

Children trust confident adults. Patients trust confident healers. Communities trust confident religious leaders.

The speaker’s certainty becomes part of the evidence, even though certainty itself proves nothing.

History contains countless people who were completely certain and completely wrong.

The mature question is not:

“How strongly do I believe this?”

It is:

What justifies my confidence?

Living Without False Closure

Rejecting false certainty does not require becoming uncertain about everything.

We can know many things with strong confidence.

Evidence shows that germs cause disease, the earth orbits the sun, and species evolve. We can also make sound moral judgments. Cruelty causes suffering. Children require protection. Patients deserve care. Birth should not determine human worth.

What we abandon is not knowledge.

We abandon the demand for absolute answers where evidence does not support them.

Some questions may remain open.

Why does anything exist?

What is the full nature of consciousness?

What happened at the earliest limits of the universe?

Does any part of personal experience continue after death?

An open question is not a failure.

It is an honest boundary around what we know.

Religion often fills such boundaries with confident stories.

A free mind can leave them open without panic.

This may feel difficult at first. A person raised with certainty may experience doubt as emptiness. The old answer disappears before a new sense of stability has formed.

That transition deserves patience.

Leaving certainty does not mean losing all direction.

It means building direction on firmer foundations:

evidence where evidence is available,

compassion where people suffer,

and humility where knowledge ends.

The Courage to Remain Uncertain

False certainty promises rest.

It says the universe is governed, suffering has meaning, death is not final, and the believer possesses the correct answer.

The promise can be deeply comforting.

But comfort obtained by closing the mind carries a price.

It may stop investigation, protect injustice, blame the sufferer, and give people authority over others. It may persuade us to defend a doctrine more carefully than the person standing before us.

Honest uncertainty offers less emotional shelter.

It offers something more reliable:

freedom from pretending.

A person who can say "I do not know" is harder to manipulate. He does not need a priest, guru, astrologer, politician, or miracle seller to fill every silence.

He can listen.

He can examine.

He can wait.

He can change his mind.

That does not make him cold.

It may make him more compassionate, because he no longer needs another person's suffering to fit a theory.

Certainty may feel sweet.

Truth is not obligated to be sweet.

Key Points

- Religious certainty is attractive because uncertainty feels emotionally painful.
- Comfort may be genuine without proving the belief that provides it.
- Quick religious explanations can burden grieving or suffering people.
- Final answers may stop investigation into preventable causes and human responsibility.
- Sacred certainty can give people unjustified authority over others.
- Science gains strength from its willingness to revise conclusions.
- "I do not know" protects against invented explanations and false authority.

Reader Reflection

Which religious certainty gives me comfort, and what evidence supports it?

Have I ever explained another person's suffering when simple compassion would have been kinder?

What question am I afraid to leave unanswered?

Transition to Chapter 4

Certainty offers relief from uncertainty, but fear gives it urgency.

Once people believe that invisible forces can punish, contaminate, curse, or protect them, ritual and authority gain power over ordinary life.

The next chapter examines how ordinary human fear is transformed into sacred duty.

Chapter 4: When Fear Becomes Holy

A baby cries through the night.

The parents check the temperature, feed him, hold him, and walk from room to room. Nothing seems to settle him. By morning, everyone is tired and worried.

Then a relative offers an explanation.

"Too many people praised him yesterday. Someone's evil eye may have fallen on him."

The words change the atmosphere.

Until that moment, the family faced uncertainty. The baby might be hungry, uncomfortable, overtired, or becoming ill. Now the fear has a name.

Someone brings chilies, salt, incense, or another familiar object. A small ritual is performed. The adults watch closely for signs that the danger has left.

Perhaps the baby eventually stops crying.

The ritual receives credit.

If he continues, the family may decide that the evil eye was stronger than expected, the ritual was incomplete, or another remedy is needed.

The belief has entered a protected circle.

This is how ordinary fear becomes sacred.

It begins with a real human emotion. A child is ill. A pregnancy ends. A business fails. Several misfortunes occur close together. A disturbing dream is followed by bad news.

People feel exposed and want an explanation.

When evidence does not provide one quickly enough, a supernatural story may enter the gap.

A spirit is angry.

A planet is unfavorable.

A ritual was neglected.

A deity is displeased.

Someone cast a curse.

The explanation may have no reliable evidence, but it offers something fear desperately wants:

a cause,

a meaning,

and something to do.

Fear Wants a Story

Uncertainty is difficult because it leaves the mind without direction.

A family may be able to tolerate a known problem better than an unknown one. Even a frightening explanation can feel more manageable than no explanation at all.

A supernatural story supplies order.

The illness is not random.

The loss is not meaningless.

The failure did not simply happen.

Someone or something caused it.

This sense of intention is emotionally powerful.

Natural explanations are often less satisfying. Illness may involve infection, genetics, environment, stress, or several causes acting together. A miscarriage may occur without anyone being responsible. A business may fail through poor planning, changing conditions, bad luck, or a combination of all three.

Reality may offer only a partial answer.

A curse feels complete.

So does karma, fate, possession, or divine punishment.

The story seems convincing because it matches the emotion. A frightening event appears to deserve a frightening cause. A tragedy feels too important to be accidental. A sudden recovery feels too fortunate to be natural.

The strength of the feeling is then mistaken for the strength of the evidence.

I have seen educated and loving families reach for these explanations during crisis. They were not foolish. They were frightened.

Fear makes an inherited story feel safer than uncertainty.

Ritual Gives Fear Something to Do

Once fear receives a supernatural explanation, ritual follows naturally.

If the problem is evil eye, something must remove it.

If a deity is displeased, an offering may restore favor.

If a planet is dangerous, a ceremony, fast, stone, or donation may reduce its influence.

If a spirit is present, someone must drive it away.

Ritual gives the family action.

That matters psychologically.

Helplessness is one of the most painful parts of fear. Doing something, even something ineffective, can produce relief. The family no longer feels passive. It has responded.

The relief may be genuine.

The family feels calmer. The ritual creates order. People gather, repeat familiar words, and regain a sense of control.

But emotional relief does not demonstrate that an invisible danger was removed.

The ritual may change how people feel without changing what caused the problem.

This distinction is easy to miss because the timing can be persuasive.

A child's fever improves after a ceremony.

A journey ends safely after an auspicious time is chosen.

A family performs a ritual, and the feared event does not occur.

The ritual receives credit.

If the feared event occurs anyway, the ritual is rarely allowed to fail clearly. Believers may say it was begun too late, performed incorrectly, or not strong enough.

The system can survive every outcome.

Success confirms it.

Failure changes the explanation.

Authority Makes Fear Respectable

A private fear may fade.

A sacred fear is harder to challenge.

If one person says, "I have a bad feeling," the family may ignore it. If a priest, guru, astrologer, healer, elder, or sacred text gives the fear religious authority, the warning carries greater weight.

The claim now has status.

It appears to come from someone with special access to hidden knowledge.

The date must be changed.

The ritual must be performed.

The offering must be made.

The healer must be consulted.

The family may not understand why.

Understanding is no longer necessary because authority has interpreted the danger.

This pattern appears in many religious traditions. The titles, clothing, rituals, and language differ, but the structure is similar. A person claims knowledge that ordinary people cannot verify, and fear increases the willingness to obey.

Some religious authorities are sincere. They believe the system they represent. Others may recognize that fear produces money, loyalty, or influence.

The effect on the frightened person can be the same.

He is no longer deciding only whether the claim is true.

He must also decide whether he can risk opposing an elder, disappointing the family, or violating a sacred custom.

Authority turns fear into duty.

Guilt Makes Obedience Feel Voluntary

Sacred fear does not always need force.

Guilt often does the work.

A person may doubt a ritual but participate because he fears being blamed if something later goes wrong.

A parent may obey an astrologer because refusing feels irresponsible.

A young couple may accept a religious remedy because no one wants to hear, after a difficult marriage, "We warned you."

The thought is simple:

What if they were right?

This question can control people who do not fully believe.

No one can guarantee that misfortune will never follow a rejected ritual. Life contains enough illness, conflict, accidents, and disappointment that something unpleasant will eventually happen.

If it does, the old warning may return.

Perhaps we should have listened.

Perhaps the ceremony mattered.

Perhaps we caused this.

The fear becomes difficult to escape because ordinary life will always supply events that can be interpreted as punishment.

The person technically remains free to refuse.

But refusal has been made emotionally dangerous.

He must carry not only the ordinary risk of life, but the imagined risk of divine anger, bad karma, family blame, or supernatural harm.

This is not fully free choice.

A decision made under an unsupported threat is still shaped by coercion, even when the threat is invisible.

When Fear Is Called Tradition

Most people do not like being described as superstitious.

They use gentler words.

Tradition.

Culture.

Faith.

Respect.

Precaution.

Ancestral wisdom.

These words may describe something valuable. A festival can create joy. A mourning ceremony can bring relatives together. A symbolic blessing can express love. Culture gives people memory, identity, and belonging.

The problem begins when fear hides inside the custom.

A practice is called tradition, but the family is afraid to stop it.

A ritual is called respect, but refusal brings guilt.

A restriction is called purity, but it humiliates someone.

A ceremony is called devotion, but a poor family feels forced to pay.

The name can conceal the mechanism.

A useful question is:

What do we believe will happen if we refuse?

If the answer involves divine anger, impurity, bad luck, supernatural punishment, or danger to the dead, fear is present whatever language surrounds it.

This distinction allows culture to survive without captivity.

A family can celebrate a ceremony because it is meaningful, not because disaster is expected if one detail is missed.

It can remember the dead without believing that their peace depends on payment.

It can preserve a symbol without giving the symbol power over reality.

Beauty does not require fear.

Fear Looks for a Target

Supernatural fear becomes more dangerous when it attaches itself to a person.

A community facing illness, death, crop failure, or repeated misfortune may search for someone to blame.

A widow is called unlucky.

A woman is accused of witchcraft.

A child is believed to carry a curse.

A person with mental illness is treated as possessed.

Someone from a lower caste is considered polluting.

The explanation gives fear a human target.

Once that happens, compassion weakens. The person is no longer seen primarily as someone needing care or protection. He or she becomes a carrier of danger.

This process appears across cultures because blame offers emotional relief. A complicated problem feels simpler when one person can be identified as its source.

The target is rarely powerful.

Fear moves downward.

It falls most easily on those with the least ability to defend themselves.

Later chapters will examine the victims more fully. Here the central point is how the process begins:

uncertainty becomes supernatural explanation,

and supernatural explanation identifies someone through whom danger is imagined to operate.

An unsupported spiritual accusation is therefore never harmless.

Ideas change how people are treated.

Why Sacred Fear Resists Failure

A belief should weaken when its predictions repeatedly fail.

Sacred fear often avoids this correction.

If prayer is unanswered, the outcome becomes God's mysterious will.

If a ritual fails, someone performed it incorrectly.

If an astrologer is wrong, another part of the chart is emphasized.

If a healer does not help, the patient's faith is blamed.

Every result can be absorbed.

A system that interprets all possible outcomes as support cannot learn from reality.

This helps explain why superstition survives in technologically advanced societies. A person may use a smartphone, travel by airplane, and work in a modern hospital while still believing that a ritual prevented bad luck.

The belief has never been required to face a fair test.

A fair test asks:

What exactly is being predicted?

How often does it occur without the ritual?

Are failures recorded?

What result would show that the claim is mistaken?

Would we accept the same explanation from another religion?

If no possible evidence can weaken a belief, the belief is not being protected by truth.

It is being protected from truth.

Breaking the Sequence

Sacred fear gains power through a recognizable sequence:

uncertainty becomes supernatural explanation,

the explanation produces ritual,

ritual brings authority,

authority creates guilt,

and guilt produces obedience.

The sequence can be interrupted at any point.

A family can pause before assigning a supernatural cause.

It can ask whether a natural explanation is more likely.

It can seek practical help before ritual.

It can question anyone claiming hidden knowledge.

It can refuse to blame a vulnerable person.

The first response need not be ridicule.

Mockery may deepen defensiveness because the believer feels personally attacked. A calmer question can be more effective:

"What evidence supports this?"

"What would show that the claim is wrong?"

"Would we accept this reasoning from another religion?"

"Are we acting because the claim is true, or because we are afraid?"

"What practical action is needed now?"

These questions return attention to reality.

The fear may not disappear immediately. A person abandoning a ritual may remain anxious. An old warning can continue echoing even after the belief has weakened.

That anxiety does not prove the warning was true.

It shows that fear was learned deeply.

Each refusal teaches the mind something new.

The journey begins at an ordinary time and ends safely.

The ritual is omitted, and no supernatural punishment follows.

The family chooses a simpler ceremony, and love remains.

Reality slowly replaces fear.

Culture After Fear

Some people worry that questioning superstition will destroy culture.

It need not.

Culture can retain music, food, art, celebration, memory, ceremony, and family connection without preserving every supernatural claim.

A ritual may remain as symbolism if participation is voluntary and no one is threatened by invisible consequences.

This requires honesty.

Adults should not tell children that a custom is "only cultural" while behaving as though disaster will follow if it is ignored. Children learn more from anxiety than from explanation.

Earlier generations lived with less scientific, medical, and psychological knowledge. They created explanations that made sense within their world.

We can understand why they believed without pretending that every inherited explanation remains true.

Respect for the past does not require submission to it.

We can keep the lamp and abandon the fear.

We can keep remembrance and abandon coercion.

We can keep community and allow people to choose.

When fear is removed, culture does not become empty.

It becomes freer.

Key Points

- Sacred fear often begins with a real event and an uncertain explanation.
- Supernatural stories give fear a cause, meaning, and apparent remedy.
- Ritual reduces helplessness, but emotional relief does not prove supernatural power.
- Religious authority can turn private anxiety into respected obligation.
- Guilt and fear of blame can make apparent choice less than fully free.
- Tradition and culture may conceal superstition when refusal is expected to bring danger.
- Sacred fear often seeks vulnerable people to blame.

- Beliefs that interpret every outcome as confirmation cannot be corrected by evidence.

Reader Reflection

Which practice in my family is followed mainly because people fear what might happen if it stops?

Have I ever obeyed a religious warning even though I did not truly believe it?

What tradition would remain meaningful if its fear were removed?

Transition to Chapter 5

Sacred fear often concerns dangers no one can see.

Miracle stories appear to offer something stronger: visible proof that supernatural powers are real.

A statue seems to drink. A patient recovers. A prophecy appears fulfilled.

The next chapter asks what such stories actually prove.

Chapter 5: Miracles: Beautiful Stories, Poor Evidence

In September 1995, a remarkable story spread across India.

Statues of the Hindu god Ganesha appeared to be drinking milk.

Devotees held spoons near the statues' trunks. The milk seemed to disappear. News traveled quickly through temples, homes, television, and telephone calls. Crowds formed. Similar reports soon appeared in other countries.

For many believers, the meaning was immediate.

The gods had acted.

A miracle was unfolding in public.

People who witnessed the event were not necessarily lying. Many were sincere, excited, and convinced by what they saw.

But sincerity does not settle what happened.

Investigators offered natural explanations. Small amounts of milk could move along the surface of a statue through capillary action and then spread, drip, or collect out of sight. The effect varied with the material and shape of the statue. It did not establish that a divine being had consumed the liquid.

Yet by the time ordinary explanations appeared, the miracle had already entered public memory.

The episode reveals how miracle stories gain power.

An unusual observation occurs.

Expectation gives it meaning.

Crowds repeat the interpretation.

Religious identity strengthens conviction.

The uncertain event becomes proof.

Miracles are persuasive not only because something unusual happened, but because people deeply want the event to mean something.

That desire deserves understanding.

It does not replace evidence.

Why Miracles Feel Like Proof

Gods, spirits, saints, and divine powers are usually invisible.

A miracle seems to make them visible.

A statue drinks.

A patient recovers.

A prophecy appears to come true.

A prayer is followed by unexpected good news.

The invisible belief suddenly seems to have left a physical mark on the world.

That is emotionally powerful.

A philosophical argument for God may be difficult to follow. A story about a dying patient who recovered after prayer can be understood in seconds.

The doctor said there was no hope.

The family prayed.

The patient survived.

Therefore, God intervened.

The story is clear, moving, and easy to remember.

Real life is rarely so simple.

The diagnosis may have been uncertain. Treatment may already have been working. The prognosis may have been misunderstood. The physician may have said that recovery was unlikely, not impossible.

Those details make the story less dramatic, so they gradually disappear.

What remains is the miracle.

An Experience Can Be Real While Its Explanation Is Wrong

People often defend a miracle by saying:

"I saw it."

“It happened in my family.”

“Thousands witnessed it.”

“The doctor could not explain it.”

Personal testimony matters. It may give us a reason to investigate.

But testimony is not always reliable enough to establish a supernatural cause.

People misperceive events.

Memory changes.

Expectation influences observation.

A story becomes simpler and more dramatic each time it is repeated.

Crowds can intensify these effects. Once several people announce that something extraordinary is happening, later observers approach the event expecting to see a miracle.

The setting matters too.

A person entering a temple, church, shrine, pilgrimage site, or healing service is already surrounded by sacred expectation. Prayer, music, ritual, authority, and the emotion of the group influence how an ambiguous event is understood.

Someone may genuinely feel warmth, trembling, peace, pressure, or a sense of presence.

A patient may genuinely improve after prayer.

A devotee may honestly believe that a statue responded.

The experience is real as an experience.

Its interpretation remains open.

That distinction is not disrespectful.

It is essential.

The Medical Miracle

Some of the most powerful miracle stories arise in hospitals.

A patient is critically ill. The family prays. Treatment continues. Against expectation, the patient improves.

The recovery is attributed to God, a saint, a guru, sacred water, or the prayers of the family.

I understand why this happens.

When someone survives, gratitude searches for a direction. Medical language may sound cold beside the words "God saved him."

But recovery alone does not prove supernatural intervention.

Medicine deals in probabilities, not perfect prediction.

A physician may say that a patient has a ten percent chance of survival. That still means some patients will live.

A treatment may begin working later than expected.

An initial diagnosis may be revised.

The body may respond in an unusual but natural way.

The family may also misunderstand what the doctor actually said. “The outlook is poor” becomes “There is no hope.” “Recovery is unlikely” becomes “The doctors said survival was impossible.”

Then the patient survives.

The story becomes stronger than the original medical facts.

Treatment may also vanish from the retelling.

The patient received surgery, antibiotics, intensive care, oxygen, chemotherapy, dialysis, rehabilitation, or months of skilled nursing.

Yet prayer receives the decisive credit.

Families are free to pray. Prayer may provide courage and emotional support.

But gratitude should not erase the work of medicine, the natural capacity of the body, or the role of probability.

A surprising recovery is evidence that a surprising recovery occurred.

It is not automatically evidence that a supernatural being caused it.

The Patients Who Did Not Recover

Miracle stories are usually told by survivors.

The failures remain quiet.

A child improves after visiting a shrine, and the family tells the story for years.

Another child visits the same shrine and dies.

That death does not become evidence against the shrine.

A patient recovers after prayer and gives public testimony.

Another patient receives equally sincere prayer and continues to decline.

The prayer itself is not questioned.

Success is remembered.

Failure is given another explanation.

The patient lacked faith.

The ritual was incomplete.

God had another plan.

Death itself was the answer.

Karma could not be changed.

These explanations protect the belief from failure.

If recovery proves divine intervention, then non-recovery should count against the claim. If every outcome can be reinterpreted as divine wisdom, the belief can never lose.

That is not a fair test.

Imagine one thousand seriously ill patients receiving prayer.

Some recover completely.

Some improve partially.

Some remain unchanged.

Some die.

If only the successful cases are publicized, prayer will appear powerful even if the overall outcomes are no different from what medicine and probability would predict.

Reliable evaluation requires the complete record.

How many people were treated?

How many recovered?

What treatment did they receive?

How many similar patients recovered without the prayer or ritual?

Were the outcomes recorded consistently?

Without these questions, a moving story remains a story.

It does not become proof.

Every Religion Has Miracles

Nearly every religion reports miracles.

Christians speak of healings, apparitions, answered prayer, and the intervention of saints.

Hindus describe sacred ash, divine dreams, healing gurus, blessings, and statues showing supernatural behavior.

Muslims tell of healings, sacred signs, answered prayers, and wonders associated with holy people or places.

Similar reports appear in Sikh, Buddhist, Jewish, Indigenous, spiritualist, and newer religious traditions.

The stories often support contradictory beliefs.

A Christian miracle may be presented as evidence that Jesus is the unique path to salvation.

A Hindu miracle may be understood as proof of a deity or guru.

A Muslim miracle may be taken as confirmation of Islam.

They cannot all establish their exclusive religious conclusions.

This leads to a useful test:

Would I accept this evidence if it supported another religion?

Believers often become skeptical when the miracle belongs to someone else.

They consider exaggeration, coincidence, error, fraud, suggestion, or natural explanation.

That same skepticism should be applied at home.

If testimony is weak evidence for another religion's miracle, it does not become strong merely because the story supports our own beliefs.

A changing standard reveals loyalty, not truth.

What Crowds Can and Cannot Show

The statement "Thousands of people saw it" sounds impressive.

Large numbers may increase confidence.

They do not automatically improve the quality of the evidence.

A thousand people can share the same expectation.

They can observe the same ambiguous event.

They can hear the same interpretation before looking for themselves.

They can influence one another.

During the Ganesha milk episode, reports from one location encouraged people elsewhere to perform the same act. Those later observers were not approaching a neutral experiment. They were trying to reproduce something they had already been told was miraculous.

A crowd is not the same as a thousand independent investigations.

Numbers matter only when the observation is clear, the reports are independent, the conditions are controlled, and alternative explanations are examined carefully.

Otherwise, repeated testimony may multiply one mistake.

Human beings do not become less suggestible merely because they gather in large numbers.

Sometimes the crowd makes independent judgment harder.

When the Claim Changes

Miracle claims often begin as clear statements about the physical world.

The statue drank milk.

The patient was instantly cured.

The prophecy named the event in advance.

The prayer changed the outcome.

These claims appear testable.

When investigation creates difficulty, the claim may change.

The miracle was symbolic.

The real healing was spiritual.

Science cannot measure divine action.

God does not perform on demand.

The event was meant to strengthen faith, not provide proof.

A symbolic interpretation may hold personal meaning.

But it is not the same claim.

If a physical event is offered as evidence, it should remain a physical claim long enough to be examined.

A statue cannot be presented as literally drinking milk and then become metaphorical when a natural explanation is found.

A cure cannot be called physical proof and later redefined as emotional healing when medical records do not support it.

This movement protects belief by making it difficult to test.

A claim cannot demand the authority of fact while avoiding the responsibilities of fact.

What Strong Evidence Would Look Like

A skeptic need not reject every unusual event in advance.

Evidence for an extraordinary event could become compelling.

Suppose a healer repeatedly restored a clearly documented physical function under controlled conditions.

Suppose independent medical teams confirmed the same result.

Suppose successful and unsuccessful cases were recorded honestly.

Suppose ordinary explanations were carefully excluded.

That would deserve serious investigation.

Or suppose a statue repeatedly consumed a measurable amount of liquid under conditions that excluded absorption, hidden channels, evaporation, and human interference.

That too would require attention.

The issue is not that science has already explained every strange event.

It has not.

The issue is that an unexplained event is not automatically a supernatural event.

"Unexplained" means that the explanation is not yet known.

It does not mean that one religion's explanation has been proved.

A careful investigation asks:

What exactly happened?

Was it documented before the story spread?

Can the event be repeated?

Were ordinary explanations excluded?

Were failed cases recorded?

Are the investigators independent?

Would the same evidence persuade us if it came from another faith?

The more extraordinary the claim, the stronger the evidence must be.

That is not hostility.

It is proportion.

When Hope Becomes Vulnerable

Miracle claims become especially serious when they influence health, money, or authority.

People who seek miracles are often frightened or desperate. They may face cancer, infertility, disability, chronic pain, grief, or an illness medicine cannot cure.

Hope weakens skepticism because disbelief feels dangerous.

“What if this is the one thing that might help?”

That question can lead families toward expensive ceremonies, unproven remedies, repeated donations, or delays in proper care.

Some healers and religious leaders may sincerely believe their claims.

Others may recognize that miracle stories bring followers, status, money, or power.

Sincerity does not remove the need for scrutiny.

When a miracle claim affects a patient's treatment or a family's finances, ask:

Are complete records available?

Are unsuccessful cases disclosed?

Can independent physicians examine the evidence?

Does the claimant benefit financially or socially?

Are patients encouraged to continue appropriate medical care?

A person with genuine healing power should welcome careful documentation.

Avoiding examination does not demonstrate holiness.

Wonder Does Not Need a Miracle

Rejecting a supernatural explanation does not require rejecting wonder.

The natural world is astonishing.

A human body develops from a single fertilized cell.

The brain produces memory, language, music, grief, imagination, and love.

Stars formed the elements that later became planets, blood, bone, and breath.

None of this becomes less beautiful because it can be studied.

Natural explanation does not empty life of meaning.

It often deepens wonder by revealing how much complexity lies beneath ordinary experience.

A patient's recovery after skilled care remains worthy of gratitude.

A sunrise does not require divine intervention to be beautiful.

A family reunion after danger does not need supernatural explanation to be precious.

The word *miracle* may be used poetically.

A parent may call a child's recovery miraculous because it felt extraordinary.

That language need not cause confusion if everyone understands that it expresses gratitude rather than scientific proof.

The problem begins when poetry is turned into evidence.

Wonder survives explanation.

Only a weak claim needs mystery to protect it.

A Story Is Not Yet Proof

Miracle stories deserve humane listening.

They often arise from fear, illness, grief, gratitude, and hope. The person telling the story may have endured something painful and may attach deep meaning to what happened.

Ridicule is unnecessary.

But respect for the person does not require agreement with the explanation.

We can listen carefully and still ask what actually occurred.

We can acknowledge the recovery while examining its medical history.

We can respect the witness while considering error, expectation, and coincidence.

We can value the emotional meaning without accepting supernatural causation.

This distinction protects compassion and truth at the same time.

Most miracle claims offer some combination of testimony, coincidence, selective memory, natural recovery, crowd influence, or an event that remains incompletely explained.

These may inspire belief.

They do not establish it.

A beautiful story can lift the heart.

Before it is allowed to govern the mind, it must face the evidence.

Key Points

- Miracle stories are persuasive because they combine mystery, hope, emotion, and religious expectation.
- A sincere experience may be real even when its supernatural interpretation is mistaken.
- Unexpected medical recovery does not by itself prove divine intervention.
- Miracle traditions usually preserve successes while giving little attention to failures.
- Contradictory religions use similar miracle stories to support incompatible beliefs.
- Crowds can multiply confidence without producing independent evidence.
- A physical claim should not become symbolic only after investigation creates difficulty.
- An unexplained event is not automatically a supernatural event.
- Claims affecting health, money, or authority require especially careful scrutiny.

Reader Reflection

Would I accept the evidence for my religion's miracle if it supported another faith?

Do I remember apparent answers to prayer more carefully than unanswered prayers?

Have I ever confused the emotional importance of an event with proof of its supernatural cause?

Transition to Chapter 6

Miracle stories remain persuasive partly because religion is often allowed to use a lower standard of evidence than other serious claims.

Faith, testimony, tradition, and personal feeling may be accepted where careful proof would normally be required.

The next chapter examines this special exemption and asks why sacred claims should not be judged by the same standards we use everywhere else.

Chapter 6: The Special Exemption

Imagine that I enter a hospital room and tell a frightened family:

“I have a cure for your loved one.”

They ask what the treatment contains.

I refuse to explain.

They ask whether it has been tested.

I say testing is unnecessary.

They ask how many patients improved.

I offer three moving stories.

They ask whether independent physicians have examined the results.

I become offended.

“Why are you questioning me? You must have faith.”

No responsible patient, family, or physician should accept such an answer.

The claim is too important.

A treatment that may influence life or death must face careful scrutiny. We would ask what was given, how it was tested, how many patients received it, what other treatments they had, what risks occurred, and whether independent investigators found the same result.

These questions would not be hostile.

They would be necessary.

Now consider a religious claim:

Prayer healed the patient.

A sacred object prevents danger.

A scripture came from God.

A prophet received divine revelation.

A ritual changes future events.

A soul survives death.

These claims are at least as important as a medical claim. They influence how people raise children, treat illness, choose marriage partners, judge behavior, spend money, and understand death.

Yet when similar questions are asked, the standards often change.

"You must have faith."

"Spiritual truth cannot be tested."

"Millions of people believe it."

"You cannot understand until you experience it."

"Some things are beyond reason."

The claim is allowed to shape human life while being protected from the examination normally expected of serious claims.

Religion has received this special exemption for so long that many people no longer notice it.

They demand evidence in medicine, engineering, law, and finance.

Then the subject becomes sacred, and faith is treated as enough.

The Rules We Already Use

Daily life depends on trust.

We trust pharmacists to dispense the correct medication, pilots to fly safely, engineers to design bridges, and historians to describe events we did not witness.

This trust is not blind.

It rests on evidence, training, accountability, and correction.

Pharmacists are licensed.

Aircraft are inspected.

Bridges follow engineering standards.

Historical claims are compared with records.

Scientific findings can be challenged and repeated.

Trust becomes reasonable because it can be checked.

We also adjust the amount of evidence to the size and consequence of the claim.

If a friend says he drank tea this morning, little evidence is needed. The claim is ordinary and carries almost no consequence.

If he says that a certain tea cured his cancer, the standard changes.

We would want medical records, the original diagnosis, treatment history, and independent evaluation.

If he claims that the tea cures every cancer, the evidence must become stronger still.

Religious claims often belong to this final category.

They concern the origin of the universe, invisible beings, divine commands, miracles, life after death, eternal punishment, and moral rules imposed on other people.

These are not small private impressions.

They are claims about reality.

The larger and more consequential the claim, the stronger the reason for accepting it should be.

Religion often asks for the opposite.

Its greatest claims are supported by its weakest standards.

Who Must Provide the Evidence?

When someone makes a claim, the responsibility to support it belongs primarily to the person making it.

This is the burden of proof.

Suppose a stranger tells you that an invisible animal lives in your home.

You cannot see, hear, touch, or detect it.

The stranger then demands that you prove the animal does not exist.

The demand is unreasonable.

Almost anything can be imagined.

An invisible spirit may live in the attic.

A god may control every coincidence.

An undetectable force may guide every decision.

If disbelief required disproving every imagined being, rational judgment would become impossible.

The person making the claim must provide a reason to accept it.

The same principle applies to religion.

A skeptic is not required to disprove every god human beings have imagined.

The believer who claims that a particular god exists, issued a revelation, performed a miracle, or established a moral command must support that claim.

Religious debate often reverses this responsibility.

"Can you prove that God does not exist?"

"Can science prove there is no afterlife?"

"Can you prove that prayer never works?"

Absolute disproof is often impossible.

But the inability to disprove a claim does not make the claim reasonable.

No one can prove that invisible spirits never move objects when people are not looking.

That does not give us a good reason to believe they do.

Possibility is not evidence.

Belief requires positive support.

Faith Cannot Choose Among Religions

Faith is often offered when evidence reaches its limit.

For many believers, faith means trust, commitment, loyalty, or hope. It may describe a deeply personal relationship with a religious tradition.

That experience can be sincere and meaningful.

But faith cannot reliably tell us which religion is true.

People use faith to reach incompatible conclusions.

A Christian may believe through faith that Jesus is the unique son of God.

A Muslim may believe that God has no son and that the Quran is the final revelation.

A Hindu devotee may trust in Krishna, Shiva, Kali, or another divine form.

A Sikh may understand the divine through a different tradition.

These positions cannot all be literally correct in the same way.

The problem is not that the believers lack sincerity.

They may be equally sincere.

They may be equally certain.

They may be equally willing to sacrifice for what they believe.

If the same method leads thoughtful people to contradictory conclusions, that method cannot determine which conclusion corresponds to reality.

Faith often preserves the religion already attached to family, culture, or conversion.

It gives emotional confidence to a conclusion.

It does not independently verify the conclusion.

Most believers already understand this when judging another faith.

A Christian does not accept a Hindu miracle merely because a Hindu has faith.

A Hindu does not accept Christian salvation claims merely because Christians believe deeply.

A Muslim does not declare every other scripture divine because its followers are sincere.

Each person recognizes that another believer's faith is not enough.

Intellectual fairness requires applying the same lesson at home.

Experience and Interpretation

Many people ground their religion in personal experience.

They feel a divine presence during prayer.

A dream appears meaningful.

A coincidence seems too precise to be accidental.

Meditation produces peace or unity.

A person grieving a death feels that the deceased is near.

These experiences can be powerful.

They may help someone endure loneliness, addiction, fear, or grief. Dismissing them casually would be unnecessary and unkind.

But an experience and its explanation are not the same thing.

A person may genuinely feel watched without anyone watching.

He may feel certain that a dream predicted an event, while forgetting hundreds of dreams that did not.

A sense of presence may arise during grief, exhaustion, meditation, neurological disturbance, or intense expectation.

The experience is real as an experience.

Its cause remains open.

Different cultures interpret similar experiences differently.

A Christian may experience Jesus.

A Hindu may experience Krishna.

A spiritualist may experience an ancestor.

Another person may interpret the same sensation psychologically.

Personal experience can reasonably support statements such as:

“I felt something profound.”

“It gave me courage.”

“It changed my life.”

It cannot, by itself, establish:

“My theological interpretation is therefore true for everyone.”

Private meaning deserves respect.

Public claims require evidence that others can examine.

When the Claim Moves

Religious claims often begin clearly.

Prayer changes physical outcomes.

God intervenes in the world.

A miracle occurred.

Scripture accurately describes creation.

A ritual prevents danger.

These claims concern reality and appear open to investigation.

When examination creates difficulty, the claim may move.

Prayer does not change events; it changes the believer.

The miracle was spiritual rather than physical.

The creation story was never intended literally.

The ritual does not prevent danger; it expresses devotion.

God acts in ways that cannot be detected.

Some reinterpretation is legitimate.

Language can be symbolic. Traditions change. A ritual may possess cultural meaning without making a supernatural claim.

The problem arises when a claim becomes symbolic only after its factual meaning fails.

A belief is first offered as evidence of supernatural action.

When evidence is requested, it retreats beyond testing.

Later, it may return as a factual claim whenever doing so is useful.

This movement makes honest evaluation impossible.

If prayer changes physical events, outcomes can be examined.

If scripture describes natural history, its claims can be compared with evidence.

If a ritual prevents misfortune, failures must count.

If God is entirely beyond observation, observations should not be presented as proof of God.

A claim cannot demand the authority of fact while avoiding the responsibilities of fact.

The essential questions are simple:

What exactly is being claimed?

What evidence supports it?

What evidence would weaken it?

If no possible observation can count against a belief, the belief is not being tested against reality.

It is being protected from reality.

Authority Is Not Evidence

Religions often rely on authority.

The scripture says it.

The prophet revealed it.

The guru knows.

The priest has studied.

The tradition is ancient.

Authority can be useful when it represents real expertise.

A physician usually understands medicine better than someone with no medical training. A historian may know an archive more thoroughly than a casual reader.

But expertise remains accountable.

A physician can be asked for evidence.

Another physician can disagree.

Treatment recommendations can change.

A historian's interpretation can be challenged by new records.

Sacred authority is often treated differently.

The source is declared correct because it is sacred, and sacred because it is believed to be correct.

"The scripture is true because God revealed it."

"How do we know God revealed it?"

"Because the scripture says so."

The reasoning moves in a circle.

A source cannot independently verify itself merely by repeating its own claim to authority.

Age does not solve the problem.

A belief may survive for centuries because it is true.

It may also survive because it is taught repeatedly, enforced socially, protected by power, or connected to identity.

Longevity shows that a belief endured.

It does not show why.

Popularity is equally weak.

Billions of people have believed incompatible religious claims.

Large numbers show social success, not factual accuracy.

Authority, tradition, and popularity may explain why a belief is widespread.

They do not establish that it is true.

When Questioning Becomes an Insult

Religious claims receive another protection when criticism is treated as an attack on believers.

Someone questions whether a scripture is historically reliable and is accused of hatred.

A person asks for evidence of a miracle and is called disrespectful.

A woman rejects an impurity rule and is accused of insulting tradition.

The discussion shifts from the truth of the claim to the feelings surrounding it.

Feelings matter.

Religion may be tied to family, ancestry, community, and painful history. Criticism can be expressed cruelly. Believers should not be mocked, threatened, or discriminated against.

But offense cannot decide truth.

People deserve dignity.

Ideas must remain open to examination.

A doctrine has no right to freedom from criticism simply because someone values it deeply.

Otherwise, any idea can become immune by attaching itself strongly enough to identity.

Respectful criticism is possible.

It represents the belief fairly.

It avoids unnecessary insult.

It admits uncertainty where uncertainty exists.

It also refuses to pretend that a harmful or unsupported claim becomes valid because questioning it causes discomfort.

A mature society must protect both personal dignity and intellectual disagreement.

Without that distinction, respect becomes silence.

Different Claims Require Different Evidence

Skepticism does not mean demanding laboratory proof for every part of life.

We do not need a randomized trial to believe that a friend is sad.

We do not need scientific measurement to appreciate a poem.

Love, grief, beauty, loyalty, and meaning are real human experiences even when they are not measured like blood pressure.

Religious arguments often blur these categories.

A person says, “God gives my life meaning.”

That may accurately describe his experience.

It is different from saying, “God created the universe and issued laws for all humanity.”

A ritual may symbolize remembrance.

That is different from claiming that the ritual changes the condition of the dead.

Prayer may calm a believer.

That is different from claiming that prayer cures infection.

The evidence must fit the claim.

Subjective experiences can be described as subjective experiences.

Historical claims require historical evidence.

Medical claims require medical evidence.

Claims about the natural world require observation and testing.

Confusion begins when emotional or symbolic meaning is used to prove an external supernatural fact.

“My faith gives me strength” may be entirely true.

“My faith therefore proves my religion is true” does not follow.

This distinction allows people to preserve personal meaning without imposing unsupported conclusions on everyone else.

One Standard for Every Religion

A fair method should not change according to whose belief is being examined.

Suppose three people make these claims:

A Christian says that Jesus appeared in a vision.

A Hindu says that a deity answered a prayer.

A Muslim says that a sacred dream conveyed divine guidance.

If sincerity, testimony, or faith is enough for one, it should be enough for all.

If psychological and natural explanations are considered for two, they should also be considered for the third.

The same principle applies to scriptures, revelations, prophecies, rituals, and healings.

This is difficult because inherited beliefs do not feel neutral.

One religion feels sacred.

The others may feel cultural, foreign, or mythological.

The difference often comes from familiarity, not evidence.

A useful discipline is to remove the familiar label.

Suppose an unknown group presents an ancient book and says it was dictated by a perfect divine being.

What evidence would you request?

Suppose another community claims that a ritual controls future events.

What would persuade you?

Suppose a stranger says his inner peace proves that his god exists.

Would you accept the conclusion?

The standard applied to an unfamiliar religion should also be applied to the religion inherited at home.

Anything less is favoritism.

Truth should not require it.

What Intellectual Fairness Requires

Religious claims should not be rejected merely because they are religious.

That would be another form of bias.

They should be examined.

A sacred text may contain historical information.

A ritual may offer psychological or communal benefit.

A religious teaching may express genuine moral wisdom.

These possibilities deserve acknowledgment.

But supernatural claims require more than usefulness, age, popularity, or sincerity.

Intellectual fairness asks us to:

define the claim clearly,

identify the evidence,

consider alternative explanations,

count failures as well as successes,

ask what would change the conclusion,

apply the same standard to competing beliefs,

and notice who may be harmed if the claim is wrong.

This method does not guarantee immediate answers.

It protects us from granting certainty too easily.

Medicine taught me that the most dangerous confidence often appears before the evidence is complete. A responsible diagnosis remains open until the findings justify it.

Belief deserves the same humility.

A claim should not gain authority because it is comforting, ancient, popular, or sacred.

Confidence should rise and fall with the quality of the evidence.

Religion has lived behind a special exemption for centuries.

Reason asks it to step outside.

Key Points

- Trust is reasonable when it rests on evidence, expertise, accountability, and correction.
- The strength of the evidence should reflect the size and consequence of the claim.
- The person making a supernatural claim carries the responsibility to support it.

- Faith cannot determine truth when different religions use it to defend incompatible beliefs.
- Personal religious experiences may be genuine without proving the interpretation attached to them.
- A factual claim should not retreat beyond testing whenever evidence creates difficulty.
- Authority, tradition, popularity, and offense do not establish factual truth.
- Religious claims should face the same standards applied to competing religions and other serious claims.

Reader Reflection

Which religious claim do I accept with less evidence than I would require from another faith?

What evidence could make me reconsider one of my central beliefs?

Do I treat criticism of an idea as criticism of my identity?

Transition to Chapter 7

Religion's special exemption is preserved most powerfully through sacred books.

Once a text is believed to contain divine truth, its authority can extend far beyond the society and century in which it was written.

The next chapter examines what happens when human beings freeze truth inside ancient words.

Chapter 7: Frozen Truth

A sacred verse is recited.

The room becomes quiet. Heads bow. The book is lifted carefully, placed on a clean cloth, touched with reverence, or carried in ceremony. No one treats it like an ordinary object.

Then someone asks:

"Who wrote these words?"

The question changes the atmosphere.

A book may be admired for its beauty, wisdom, history, or emotional power. But once it is called divine, a different expectation appears. The text is no longer simply read. It is obeyed, defended, and protected from ordinary criticism.

Growing up in a Hindu family, I understood the force of sacred words before I understood their history. Their power came through sound, rhythm, ritual, and the seriousness of the adults around me. Long before I could examine what a verse meant, I knew that it belonged to a category above ordinary speech.

That feeling was real.

It was not evidence that a god had written the words.

Every sacred text entered the world through human beings. Someone spoke, remembered, wrote, copied, selected, translated, interpreted, and preserved it.

The traditions differ greatly. Some religions emphasize a single book. Others rely on collections of texts, commentaries, oral traditions, teachers, or layered forms of authority. Not every believer reads scripture literally, and not every religious community treats every passage in the same way.

Still, one problem appears whenever a text is declared perfect:

human words are given the power to end human questioning.

A book can be beautiful and still be wrong.

How a Human Text Becomes Sacred

Most believers encounter scripture first through trust.

Parents, grandparents, clergy, teachers, and communities tell the child that the book is holy. It may be described as revealed by God, spoken through a prophet, heard by ancient sages, protected from error, or preserved through divine guidance.

The child does not begin by studying authorship, historical context, translation, or canon formation.

He begins with reverence.

The book is sacred because trusted adults treat it as sacred.

Ritual strengthens the conclusion. The text is recited on important occasions. It is displayed with respect. It may be touched, kissed, wrapped, carried, memorized, or placed above other books.

The emotional lesson comes before the intellectual examination.

By the time questions arise, the central conclusion may already feel settled.

This is not unique to one religion. Communities differ in detail, but the pattern is familiar: sacred authority is inherited before it is investigated.

The feeling of holiness then becomes part of the evidence.

A believer may say:

“I feel something when I hear these words.”

“The language is too profound to be human.”

“This book transformed my life.”

“My family has trusted it for generations.”

These statements may reveal the power of the text.

They do not establish its divine origin.

Great literature, music, and poetry can also transform a life. Emotional depth shows what a work does to the reader. It does not prove who ultimately authored it.

Every Scripture Has a History

Sacred books do not arrive outside history.

They emerge in particular languages, places, conflicts, and communities. They are shaped by the concerns of the people who preserve them.

Texts may pass through oral transmission. Different versions may circulate. Editors may combine traditions. Communities may debate which writings deserve authority. Copyists, translators, commentators, and religious institutions may influence how the words are understood.

These processes are not evidence of corruption in some crude sense.

They are what human history looks like.

A religious believer may respond that divine revelation worked through these human processes. That is possible as a theological interpretation.

It is not demonstrated merely by the existence of the text.

The book cannot prove its own divine origin by claiming that it came from God.

"The scripture is true because God revealed it."

"How do we know God revealed it?"

"Because the scripture says so."

This reasoning moves in a circle.

A claim repeated inside the source making the claim is not independent verification.

Ancient writers may have been completely sincere. They may have believed that dreams, visions, voices, revelations, or mystical experiences came from a divine source.

Sincerity helps explain why a text was written.

It does not prove that a god was speaking through the writer.

Many Books, Many Final Truths

Humanity does not possess one universally accepted sacred book.

It possesses many.

The Bible, Quran, Vedas, Upanishads, Bhagavad Gita, Guru Granth Sahib, Buddhist canons, Jain texts, Book of Mormon, and many other writings are treated with devotion by different communities.

Their followers may describe them as revealed, inspired, perfect, final, or spiritually authoritative.

These texts do not give one consistent account of reality.

They differ about God, creation, revelation, salvation, rebirth, sin, ritual, purity, marriage, sexuality, death, heaven, hell, and the path to liberation.

A Christian may consider the Bible uniquely authoritative.

A Muslim may regard the Quran as the final revelation.

A Hindu may approach several sacred traditions rather than one exclusive text.

A Sikh may revere the Guru Granth Sahib in a way that differs from all of these.

Each community has reasons for trusting its own scripture.

The reasons often resemble one another:

It has guided millions.

It contains great wisdom.

Its prophecies came true.

Its language is extraordinary.

It changed my life.

It has survived for centuries.

I feel the divine when I read it.

These claims show that a text has been influential.

They do not distinguish divine revelation from powerful human literature.

Most believers already understand this when considering another religion.

A Christian does not accept the Quran as divine merely because Muslims revere it. A Muslim does not accept the divinity of the Bible merely because Christians find it spiritually powerful. A Hindu does not regard every sacred book as a final revelation simply because its followers are sincere.

Reverence is not transferable.

That should make us examine the scripture inherited at home with the same care.

The existence of many conflicting final revelations should reduce certainty.

Instead, each community often assumes that everyone else misunderstood God.

The World Behind the Words

Ancient writers lived in worlds very different from ours.

They did not possess modern knowledge of astronomy, geology, genetics, microbiology, psychology, or human development. Their societies often accepted hierarchy, patriarchy, harsh punishment, tribal loyalty, conquest, and rigid family authority.

Sacred writings naturally reflect those worlds.

They contain compassion, poetry, moral insight, and profound observations about suffering and human behavior.

They also preserve assumptions that many modern believers no longer accept.

This mixture is understandable if the texts are human.

It becomes difficult when the texts are declared morally and factually perfect.

A perfect revelation intended for all humanity should not require centuries of reinterpretation before later generations discover that women are equal, that caste humiliation is wrong, that slavery is indefensible, or that unbelievers possess the same human dignity as believers.

Believers often say that the text was misunderstood.

Sometimes that may be true. Language can be ambiguous. Passages can be taken out of context. Religious traditions contain internal disagreement and competing moral themes.

But repeated misunderstanding across centuries creates another problem.

Why would a perfect god communicate essential truth in language so easily used to defend domination?

Why should ordinary people need specialists to explain that the apparent meaning was never the true meaning?

The historical explanation is simpler.

The text contains human insight and human limitation because human beings produced and preserved it.

Interpretation Keeps the Book Alive

No sacred book interprets itself.

Readers interpret it.

They decide which passages are literal, which are symbolic, which apply only to the past, which express universal truth, and which are best left unmentioned.

This work is unavoidable.

Ancient language may be unclear. Passages may conflict. New knowledge may make old explanations difficult to defend. Moral development may expose harms that earlier readers accepted.

Interpretation allows religion to adapt.

A creation story becomes symbolic after science challenges its literal meaning.

A social rule becomes "historical" after society rejects the inequality it supported.

A harsh command is placed in context.

A difficult passage is retranslated or explained through a gentler tradition.

This flexibility can reduce harm.

Religious reformers have often used interpretation to support equality, compassion, peace, and social change. Their work may be admirable and necessary.

But interpretation reveals something important.

The moral improvement comes from the reader.

The text remains the same.

People change.

A believer committed to equality searches for an interpretation compatible with equality. A believer committed to hierarchy may find support for hierarchy. A peaceful reader emphasizes compassion. A militant reader emphasizes punishment or conquest.

The same scripture can be used to support opposing conclusions because readers bring values to the text.

This is why sacred books rarely settle religious disagreement.

Different believers quote different passages, follow different commentaries, and insist that their interpretation reflects the real intention of God.

The phrase "Scripture clearly says" often means:

"This is the interpretation my community taught me to regard as clear."

Interpretation may rescue a text from harmful use.

It does not prove that the text was perfect from the beginning.

The Moral Judgment Comes From the Reader

Believers often say that scripture is the source of morality.

In practice, modern believers rarely follow every moral rule found in their sacred traditions.

They select.

They emphasize some teachings, reinterpret others, and quietly leave many behind.

This selection is usually guided by moral judgment.

A compassionate teaching feels worthy of preservation.

A cruel punishment is explained as historical.

A passage supporting equality is emphasized.

A passage supporting hierarchy is softened or rejected.

The reader is already using a moral standard.

That standard may come from empathy, experience, reason, human rights, social reform, or the testimony of people harmed by old rules.

When scripture agrees with these values, believers may say the scripture produced them.

When scripture conflicts with them, interpretation brings the book into line.

This does not mean sacred literature contains no wisdom.

It often does.

Ancient texts may offer profound insights about generosity, restraint, suffering, courage, pride, and compassion.

But wisdom should be accepted because it is wise.

A harmful command should be rejected because it is harmful.

Its location in a sacred book does not reverse its moral character.

If we must first use conscience to decide which parts of scripture deserve obedience, then conscience has already been placed above scripture.

When the Book Becomes an Internal Judge

Sacred texts do not influence only public rules.

They can enter the private mind.

A child learns through scripture and religious teaching which thoughts are sinful, which desires are impure, which doubts are dangerous, and which authorities deserve obedience.

A normal feeling may become temptation.

Curiosity may become pride.

Sexual development may become shame.

Doubt may become rebellion.

Suffering may become punishment.

The book becomes more than literature.

It becomes an internal judge.

The individual begins monitoring himself through sacred categories. He may feel guilty for thoughts he did not choose or ashamed of a body that has done nothing wrong.

This influence becomes especially powerful when disagreement with the text is presented as disagreement with God.

The human authors disappear.

The historical setting disappears.

The interpretation of a religious authority becomes divine command.

The person may then protect the book more carefully than he protects himself.

He may silence his own judgment, deny his experience, or accept humiliation because the alternative feels spiritually dangerous.

A book should help human beings understand life.

Human beings should not be required to injure themselves in order to protect a book.

The Trap of Perfect Authority

Declaring a text perfect changes the way it is examined.

If the book contains an apparent contradiction, the reader must resolve it.

If it conflicts with evidence, the passage must be reinterpreted.

If it supports a harmful practice, the believer must discover a hidden, more humane meaning.

If a prophecy appears to fail, the interpretation must change.

One possibility is excluded before the investigation begins:

the book may simply be mistaken.

This is not open inquiry.

The conclusion has already been protected.

A person may study scripture for decades without ever allowing the most basic question:

What if this is a collection of human writings rather than a perfect revelation?

Once that possibility is admitted, many difficulties become easier to understand.

Contradictions may reflect different writers or traditions.

Scientific errors may reflect limited ancient knowledge.

Harsh rules may reflect historical societies.

Changes in doctrine may reflect changing communities.

Competing interpretations may reflect ambiguous human language.

The human explanation requires no elaborate rescue.

It fits what we already know about how texts and institutions develop.

A believer may still find spiritual meaning in scripture.

Meaning does not require perfection.

Reading Without Kneeling

Sacred literature deserves serious reading.

It has shaped civilizations, laws, art, music, language, family life, and political power. Anyone who wants to understand history must understand the texts people treated as sacred.

But serious reading is not the same as submission.

We can ask:

Who wrote or compiled this?

When and where did it emerge?

What problem was the passage addressing?

Whose interests did it serve?

How did later communities change its meaning?

Who benefited from the interpretation?

Who was harmed?

We can admire poetry without accepting cosmology.

We can preserve wisdom without preserving hierarchy.

We can understand an ancient rule without imposing it on modern people.

We can recognize that the same book helped some and harmed others.

This approach does not destroy sacred literature.

It frees the text from the impossible burden of perfection.

The book no longer needs to explain every discovery, govern every century, or speak with one voice about every moral question.

It can be examined as part of humanity's long struggle to understand existence.

For some readers, this shift may feel like loss.

The perfect book becomes a human book.

Certainty becomes complexity.

But something valuable is gained.

The reader no longer has to defend every passage.

He can learn without obeying.

He can disagree without guilt.

He can allow evidence, conscience, and compassion to correct the text.

Reading without kneeling does not mean reading without respect.

It means respect no longer requires surrender.

Truth Must Remain Open

Knowledge grows because no book receives the final word.

New evidence changes old conclusions.

Wider moral concern exposes old injustice.

People once excluded from interpretation challenge the rules used against them.

A frozen truth cannot respond.

It remains fixed while human understanding expands.

Religious communities often survive by changing around their scriptures while continuing to call the scriptures unchanged.

The moral and intellectual work is being done by people.

We should acknowledge that honestly.

The question is not whether an ancient book contains anything valuable.

Many do.

The question is whether any book should be permitted to end inquiry.

No text should decide a scientific question regardless of evidence.

No passage should justify humiliation regardless of harm.

No claim of revelation should remove the reader's responsibility to think.

The moment a book is placed beyond correction, error can become sacred.

A free mind does not need to burn the book or fear it.

It opens the book, reads carefully, preserves what is wise, rejects what is harmful, and remembers that every word written in history must answer to knowledge and morality beyond its pages.

A book can be beautiful and still be wrong.

Recognizing that is not disrespect.

It is honest reading.

Key Points

- Sacred books entered history through human language, transmission, selection, translation, and interpretation.
- A text cannot prove its own divine origin simply by claiming divine authority.
- Different religions use similar reasons to defend conflicting scriptures as final truth.
- Sacred writings contain both insight and the limitations of the societies that produced them.
- Interpretation allows religion to adapt, but it also reveals the moral judgment brought by the reader.
- Modern believers often evaluate scripture using values learned beyond scripture.
- A sacred book can become an internal judge that teaches people to distrust their own questions and experiences.
- Ancient texts can be respected without being treated as perfect or beyond correction.

Reader Reflection

Which teachings in my religious tradition do I already reinterpret or quietly reject?

What moral standard do I use when deciding which passages deserve obedience?

Would I regard another religion's scripture as divine on the same evidence used for my own?

Transition to Chapter 8

A sacred text may remain unchanged, but reality does not.

The earth moves, species evolve, infections spread through natural mechanisms, and the brain follows biological processes whether scripture anticipated them or not.

The next chapter examines what happens when observation and discovery collide with sacred explanations.

Chapter 8: Reality Does Not Bend to Scripture

The earth moved before anyone believed it.

It circled the sun while priests preached, scholars argued, and sacred passages were interpreted as though the heavens had been arranged around humanity.

The planets followed their paths without waiting for permission from scripture.

Reality did not change when people misunderstood it.

That is the central lesson of science.

A belief may be ancient, comforting, widely accepted, and protected by powerful institutions. None of these can make nature behave differently.

The difficulty begins when a religious explanation is treated not as one human interpretation, but as a truth that evidence is forbidden to correct.

Then a telescope becomes dangerous.

A fossil becomes threatening.

A disease becomes a moral judgment.

A scientific discovery appears to challenge not only an idea, but God.

The conflict is unnecessary if religion remains a private source of meaning. It becomes unavoidable when sacred authority claims the right to decide how the natural world works.

Reality does not recognize that authority.

When the Sky Refused to Cooperate

For most of human history, the earth appeared motionless.

The sun seemed to rise, travel across the sky, and disappear at night. The stars moved above a world that felt fixed beneath human feet.

A universe centered on the earth seemed natural.

Religious and philosophical traditions gave that appearance deeper meaning. Humanity occupied a privileged place. The heavens appeared arranged around us.

Then observation began to disturb the picture.

Copernicus proposed a sun-centered model. Galileo later used the telescope to examine the sky in ways earlier generations could not.

He saw that the moon was not a smooth and perfect heavenly sphere. Jupiter had moons of its own. Venus displayed phases more consistent with movement around the sun.

The old picture could no longer contain everything people were seeing.

The history of Galileo is more complicated than the popular story of one heroic scientist opposed by every religious believer.

Church politics, personal conflict, scientific uncertainty, philosophical tradition, and disputes over interpretation all played roles.

Many religious individuals contributed to scientific learning. Not every cleric rejected observation. Galileo himself remained religious.

These qualifications matter.

But one fact remains.

Religious authority helped define what could be publicly taught as true. Galileo was required to abandon his defense of the earth's motion and spent the remainder of his life under house arrest.

The earth continued moving.

Authority could silence a man.

It could not stop a planet.

That is why the Galileo episode remains important. It shows what happens when an institution believes that sacred interpretation should limit investigation.

The error was not that ancient scripture failed to contain modern astronomy. No ancient text could be expected to anticipate every later discovery.

The error was granting scripture authority over observation.

Evidence Does Not Commit Blasphemy

When a sacred conclusion has been protected for generations, contrary evidence can feel morally offensive.

The evidence is not merely called mistaken.

It is called dangerous.

A scientist becomes arrogant.

A teacher becomes corrupting.

A question becomes blasphemous.

This confuses two very different things.

A claim may threaten a doctrine.

It does not threaten truth.

If the doctrine is correct, careful observation should strengthen it. If observation repeatedly contradicts it, defending the doctrine more aggressively does not improve its accuracy.

The proper response is further investigation.

Perhaps the observation is mistaken.

Perhaps the method is flawed.

Perhaps the interpretation of the evidence is premature.

Or perhaps the religious conclusion is wrong.

All possibilities must remain open.

Sacred authority often tries to remove the final possibility before the inquiry begins.

That is the moment reason is weakened.

The question is no longer, "What does the evidence show?"

It becomes, "How can the evidence be made to fit what we already believe?"

Science works in the opposite direction.

The conclusion must answer to the evidence.

Life Without Special Creation

The history of life produced another collision between inherited belief and observation.

Many religious traditions describe living beings as the result of intention. Species are created in recognizable forms. Humanity occupies a special place. Life reflects a divine plan.

Evolution offered a different explanation.

Living populations vary.

Some traits affect survival and reproduction.

Those traits may become more common over generations.

Species change, divide, and disappear.

Human beings belong within this history of life rather than standing outside it.

The idea was disturbing because it challenged more than a literal creation story.

It challenged human exceptionalism.

We are related to other animals.

Our bodies carry evidence of common ancestry.

Human beings did not appear separately in their present form.

The evidence for evolution does not depend on one fossil, one experiment, or one scientist.

It comes from many fields:

genetics,

comparative anatomy,

fossils,

biogeography,

embryology,

and directly observed evolutionary change.

Independent lines of evidence converge on the same broad conclusion.

Many religious people now accept evolution. Some interpret creation stories symbolically. Others believe that God guided the evolutionary process.

These positions are more compatible with science than denying evolution.

They also reveal something important.

The scientific understanding came from examining nature.

The religious interpretation changed afterward.

Before evolution became strongly established, many believers treated creation accounts as descriptions of what physically happened. After the evidence became difficult to deny, those accounts were increasingly described as poetry, symbolism, or theology.

Reinterpretation may preserve faith.

It should not be confused with scientific prediction.

Scripture did not reveal common ancestry, natural selection, genetics, or the long history of life.

Science discovered them by studying organisms, rocks, fossils, and heredity.

A Gap Is Not Evidence for God

Scientific knowledge is incomplete.

Important questions remain about the origin of life, the earliest universe, consciousness, and many details of evolution.

Religious arguments often enter these gaps.

Science cannot yet explain this fully.

Therefore, God must be the explanation.

The conclusion does not follow.

A gap in knowledge is evidence that knowledge is incomplete.

It is not evidence for a specific supernatural cause.

History should make us cautious.

Lightning once appeared to require divine action.

Disease appeared to require spirits or punishment.

The movement of planets appeared to require heavenly purpose.

Again and again, natural investigation produced better explanations.

The lesson is not that science will certainly answer every question.

The lesson is that ignorance should not be filled with confidence.

"I do not know" is more honest than "therefore God did it."

Otherwise, God becomes the name given to every unanswered question.

As knowledge expands, the divine explanation retreats.

This does not strengthen faith.

It ties faith to ignorance.

Disease Without Punishment or Possession

Before modern medicine, illness invited supernatural explanation.

A seizure could be called possession.

An epidemic could be described as divine anger.

Infertility might be treated as punishment.

Mental disturbance could be attributed to spirits.

Disability might be connected to sin, curse, or karma.

These explanations were understandable in societies without microbiology, genetics, neurology, or psychiatry.

They were also dangerous.

A supernatural diagnosis directs attention toward ritual, confession, exorcism, sacrifice, or blame.

A natural diagnosis directs attention toward the body, environment, infection, injury, and treatment.

The difference can determine whether a person receives care.

The development of germ theory transformed medicine because it identified mechanisms that could be observed and acted upon.

Clean water reduced disease.

Sterile technique reduced infection.

Vaccination prevented illness.

Antibiotics treated susceptible bacteria.

Disease no longer needed to be interpreted as a moral message.

This was not merely an intellectual improvement.

It was a moral one.

A child with epilepsy needs neurological care, not accusation.

A person with schizophrenia needs evaluation and treatment, not punishment for possession.

A woman after miscarriage needs compassion and medical care, not blame.

The movement from supernatural explanation to natural mechanism often protects dignity as well as health.

Medicine remains imperfect.

Doctors make mistakes.

Treatments fail.

Some conditions remain unexplained.

These limitations do not establish karma, curse, or possession.

They show where further investigation is needed.

The honest medical answer may be incomplete.

An unsupported religious answer is not made better by sounding complete.

The Retreat of the Supernatural

The same pattern appears repeatedly.

Thunder becomes atmospheric electricity.

Planetary motion becomes gravity.

Infection becomes microbiology.

Inheritance becomes genetics.

Mental disturbance becomes an area for neurological, psychological, and social investigation.

As natural mechanisms become understood, supernatural explanations lose explanatory work.

Religious believers may continue saying that God stands behind the mechanism.

God works through gravity.

God guides evolution.

God heals through medicine.

These claims may hold personal meaning.

They do not improve the explanation.

Adding divine intention does not make an orbit easier to calculate.

It does not help identify a gene.

It does not show which antibiotic will treat an infection.

The scientific explanation works without the supernatural addition.

The addition may remain meaningful to the believer, but it is no longer doing the work of discovery.

This distinction matters.

A symbolic interpretation may coexist with science.

A factual claim about the natural world must compete with evidence.

When believers say that science and religion never conflict, they often mean that religion has been reinterpreted until it no longer makes testable claims.

That may reduce conflict.

It also reveals which method supplied the knowledge.

Religious People Have Also Supported Science

The history of science and religion is not a simple story of constant warfare.

Religious individuals have made important scientific contributions. Religious institutions have sometimes supported education, preserved texts, funded scholarship, and established hospitals and universities.

Many scientists have been religious.

These facts should be acknowledged.

They do not remove the central problem.

A religious person can practice excellent science when he allows evidence to correct his conclusions.

A religious institution can support science when it does not place doctrine above investigation.

The conflict is not between every believer and every scientist.

It is between two methods of reaching truth.

One begins with claims that must remain sacred.

The other permits every conclusion to be challenged.

When religious belief accepts correction, conflict may be limited.

When doctrine claims final authority over nature, conflict becomes inevitable.

The issue is not the personal faith of the scientist.

The issue is whether evidence is allowed to win.

Science Accepted Selectively

Many people trust science where it is useful and resist it where it threatens identity.

They use smartphones, air travel, surgery, antibiotics, and satellite navigation.

They depend on physics, chemistry, engineering, and medicine.

Yet some reject evolution, consult astrology, defend possession as an explanation for mental illness, or treat sacred creation stories as literal history.

This is not always conscious hypocrisy.

Human beings compartmentalize.

Scientific reasoning governs professional life.

Religious belief governs family, death, morality, and identity.

A person may accept scientific methods in the laboratory while refusing to apply them to prayer or miracles.

Religious institutions can do the same.

A hospital may use advanced medical technology while spiritual explanations are still offered for suffering.

A technically educated family may demand the best treatment and then ask an astrologer to select the date.

The benefits of science are accepted.

The discipline of science is limited.

But the method does not become unfair when it approaches a sacred claim.

It remains the same:

observe,

define,

test,

compare,

and correct.

The method is resisted because the conclusion may not survive it.

Chapter 9 will examine this divided mind more fully. Here the essential point is simpler:

science cannot be treated as reliable when it confirms our needs and hostile when it challenges our beliefs.

Why Changing One's Mind Is a Strength

Religion often criticizes science because scientific conclusions change.

Recommendations are revised.

Theories become more precise.

Experts disagree.

New evidence replaces earlier understanding.

To someone seeking certainty, this may look like weakness.

It is one of science's greatest strengths.

Science changes because it is answerable to evidence.

A responsible physician changes a diagnosis when new findings appear.

A researcher abandons a hypothesis when repeated results contradict it.

A theory is refined when a better explanation accounts for the evidence.

Correction is not proof that the method has failed.

It is evidence that the method can expose error.

This does not mean every scientific conclusion is equally uncertain.

Some findings are supported by overwhelming evidence.

The earth will not return to the center of the solar system.

Germ theory will not be replaced by demonic possession.

Common ancestry will not disappear because one fossil is reclassified.

Scientific knowledge can be highly reliable while remaining open in principle to correction.

Religion often treats stability as proof.

But a belief can remain unchanged because it is true or because change is forbidden.

Permanence alone does not distinguish them.

A method that cannot admit error may appear confident.

It is not therefore trustworthy.

Science Also Requires Moral Judgment

Science does not answer every human question.

It can describe consequences, mechanisms, and probabilities.

It cannot determine every moral value by measurement alone.

Science can show that an action causes suffering.

It cannot, by itself, command us to care about suffering.

Ethics requires empathy, reasoning, experience, and moral commitment.

Science can also be misused.

Scientific knowledge has contributed to weapons, environmental damage, surveillance, and oppressive ideologies. Scientists can be biased, dishonest, ambitious, or cruel.

These failures matter.

They do not show that evidence is unnecessary.

They show that knowledge must be joined with ethics and accountability.

Science is not a substitute for morality.

Religion is not made factually correct by science's moral limitations.

Different questions require different tools.

Evidence helps us understand what is true about the world.

Moral reasoning helps us decide how human beings should act within it.

Neither should be allowed to ignore the other.

Reality Has the Final Word

Scripture may inspire, comfort, and preserve cultural memory.

It may contain wisdom and beautiful language.

It cannot decide how the natural world works.

The age of the earth is not determined by counting generations in an ancient text.

The origin of species is not settled by a creation story.

Disease does not become punishment because a religious authority says so.

Mental illness does not become possession because a family believes it.

Claims about reality must answer to reality.

Many believers recognize this and reinterpret their traditions when evidence becomes overwhelming.

That position is more reasonable than forcing evidence to obey scripture.

But the direction of correction should be acknowledged.

The telescope did not confirm the old universe.

Fossils and genes did not confirm separate creation.

Germ theory did not confirm divine punishment.

The evidence corrected the inherited explanation.

Religion adapted.

There is no shame in ancient people having incomplete knowledge.

Every generation begins with limits.

The problem arises when those limits are made sacred and imposed on later generations.

We honor the past by learning beyond it.

The universe is under no obligation to resemble an ancient story.

Life is under no obligation to preserve human pride.

The body is under no obligation to follow theology.

Reality does not become disrespectful when it contradicts belief.

It simply remains real.

Key Points

- Sacred authority becomes dangerous when it claims the right to limit observation.
- Galileo's history was complex, but his case shows that institutions can punish people without changing reality.
- Evolution is supported by converging evidence from many independent fields.
- A gap in scientific knowledge is not evidence for a supernatural explanation.
- Natural explanations of disease have replaced punishment and possession with mechanisms that can guide care.
- Religious individuals have contributed to science, but conflict arises when doctrine is placed above evidence.
- Science is trustworthy not because scientists are perfect, but because its conclusions remain open to correction.
- Claims about the natural world must ultimately be judged by evidence from the natural world.

Reader Reflection

Which scientific findings do I accept in practice but resist when they challenge my religious beliefs?

Has my interpretation of scripture changed only after evidence made an earlier interpretation difficult to defend?

What should have final authority when sacred tradition and observable reality conflict?

Transition to Chapter 9

Evidence has corrected many sacred explanations, yet education alone does not free the mind.

A person may understand evolution, use modern medicine, and work with advanced technology while preserving astrology, miracle claims, ritual fear, or inherited certainty in another part of life.

The next chapter examines this divided mind and asks why intelligence does not always produce intellectual freedom.

Chapter 9: An Educated Mind, Still in Chains

A civil engineer spends his working life calculating load, stress, pressure, and structural safety.

He would never approve a bridge because an astrologer selected an auspicious day. He expects measurements, tested materials, independent inspection, and clear responsibility.

Then his daughter plans to marry.

Birth charts are collected. The exact times of birth are checked. A religious specialist compares planetary positions and warns that the match may be unfavorable.

The engineer listens carefully.

The same man who would reject astrology in the design office may allow it to influence one of the most important decisions in his daughter's life.

This contradiction is common.

Doctors consult horoscopes.

Scientists perform rituals to prevent bad luck.

Professors fear the evil eye.

Lawyers who demand evidence in court accept miracle stories at home.

Business leaders wait for an auspicious hour before opening a new office.

These people are not unintelligent.

Many are disciplined, accomplished, and highly respected.

Their difficulty is not the absence of knowledge.

It is the division of the mind.

Evidence governs one room.

Inherited belief governs another.

The door between them remains closed.

Education Is Not Intellectual Freedom

Education can enlarge the mind.

It teaches mathematics, science, medicine, engineering, history, law, literature, and technology. It gives people skills, language, professional competence, and access to ideas beyond the world of childhood.

But education does not automatically teach a person to examine every belief by the same standard.

A student may learn the scientific method for an examination without applying it to prayer, miracles, astrology, or scripture.

He may understand evolution in the classroom and defend special creation at home.

He may recognize infection in the clinic and still participate in a ritual to remove supernatural danger.

The knowledge is present.

Its application is restricted.

Many educational systems reward memory, obedience, and performance more than independent thought. Students learn what answer will receive a good mark. They may become excellent at solving assigned problems without ever asking why certain sacred claims are excluded from examination.

A degree proves that someone completed a course of study.

It does not prove that he has questioned his deepest assumptions.

It does not show that he can resist family pressure, revise an inherited belief, or apply skepticism to ideas connected with identity and fear.

During years of medical practice, I have known highly educated people whose professional judgment was excellent and whose private decisions were still shaped by superstition.

Their education had changed what they knew.

It had not yet changed what they were willing to question.

The Mind With Two Rooms

Psychologists use the term *compartmentalization* to describe the separation of beliefs or experiences that might conflict if brought together.

In daily life, this can sometimes be useful. A person may temporarily set aside a family worry so that he can function at work.

In matters of belief, however, compartmentalization can protect contradiction.

In one room, a person accepts that diseases have biological causes.

In another, he believes illness may come from curse, punishment, or evil eye.

In one room, he trusts astronomy.

In another, planetary positions at birth determine marriage and fortune.

In one room, he demands evidence.

In another, faith is treated as evidence.

The two systems operate under different rules.

The person may never place them side by side.

Medical knowledge belongs to the hospital.

Religious explanation appears during family crisis.

Scientific reasoning belongs to professional life.

Sacred belief belongs to childhood, death, morality, and belonging.

Because the beliefs occupy different emotional territories, the contradiction may not feel like a contradiction.

The engineer does not say, “I reject planetary influence in physics but accept it in marriage.”

He says, “Engineering is science. Marriage is tradition.”

The change in label protects the claim.

Yet the consequences remain real.

A marriage may be prevented.

Treatment may be delayed.

Money may be spent.

A child may be frightened.

Calling a belief cultural does not remove its effect on human life.

Professional Skepticism, Private Surrender

Professional life forces people to become skeptical.

A physician who accepts every claim may harm patients.

An engineer who trusts intuition without calculation may endanger lives.

A judge who convicts without evidence destroys justice.

A business owner who believes every promise soon loses money.

These professions train people to ask:

What supports the claim?

What alternative explanation exists?

Who is responsible?

What could go wrong?

Can the result be checked?

Yet the same questions may disappear in religious life.

A sacred book is accepted because the family calls it divine.

A prayer is believed to have changed an outcome because one event followed another.

An astrologer is trusted without demonstrating reliable predictive ability.

The professional standard is suspended.

This does not happen because religious believers suddenly lose intelligence.

Religious belief is not experienced as an ordinary technical claim. It is attached to parents, ceremonies, grief, hope, death, and identity.

A diagnosis can be revised without threatening family memory.

A miracle connected to a beloved parent feels different.

A professor may challenge an academic theory without emotional difficulty. Questioning a verse recited by his mother may feel like an act of disloyalty.

That emotional difference deserves understanding.

But the beliefs closest to identity require more care, not less.

They possess the greatest power to escape examination and the greatest power to influence children, health, marriage, and moral judgment.

Professional skepticism should not stop at the front door.

Why Intelligence May Defend Error

We often assume that an intelligent person is less vulnerable to irrational belief.

Sometimes that is true.

Education can improve understanding of probability, expose contradictions, and weaken literal interpretations of religious stories.

But intelligence also gives people more skillful ways to defend what they already believe.

A contradiction receives a complicated explanation.

A failed prophecy is reinterpreted.

A miracle is protected with philosophical language.

A scientific discovery is absorbed into theology after it becomes undeniable.

The reasoning may sound sophisticated while the central belief is never placed at genuine risk.

This is rationalization.

The mind is not asking, "Is this true?"

It is asking, "How can I continue believing this?"

Religious belief often begins long before the arguments used to defend it.

A child learns the conclusion through family and ritual.

Years later, the educated adult discovers theology and philosophy that protect what was already emotionally established.

The later argument may be intelligent.

Its purpose may still be defense.

This weakness is not limited to religion.

Intelligent people rationalize political loyalty, financial mistakes, personal relationships, and moral failures. Education does not remove the human desire to protect identity.

Religion gives that desire unusual protection because the conclusion is sacred.

To question it may feel sinful, disloyal, or frightening.

The most revealing question is therefore not:

"How many arguments can I give for my belief?"

It is:

What evidence would make me change my mind?

If the honest answer is "nothing," the person is not investigating.

He is defending.

The Social Cost of Honesty

Belief survives not only because people find it convincing.

It survives because disbelief can be costly.

A young adult may question religion privately but continue attending ceremonies to avoid hurting his parents.

A woman may reject astrology but allow it to influence marriage discussions because refusal would create family conflict.

A physician may avoid challenging superstition among relatives because professional authority does not protect him from emotional pressure at home.

The educated person may know how to argue.

He may choose silence because he understands what speaking could cost.

Parents may experience a child's disbelief as rejection of the family.

A spouse may fear that the marriage has lost its moral foundation.

Relatives may worry about reputation, community reaction, or the future of religious traditions.

The doubter is asked to carry two burdens:

the intellectual burden of a belief he no longer accepts,

and the emotional burden of protecting others from his honesty.

Some compromise is unavoidable in family life.

A person may attend a ceremony out of affection. He may remain silent about a harmless symbol because disagreement would serve no useful purpose.

That may be kindness.

But silence becomes morally different when it allows harm.

A medical treatment is postponed.

A child is terrified.

A young couple is separated by astrology.

A woman is excluded or humiliated.

Not every disagreement requires confrontation.

Wisdom includes patience, timing, and sensitivity.

But family peace should not mean that the most fearful person always controls everyone else.

At some point, education must become courage.

Success Can Hide Captivity

Professional success creates an aura of authority.

A respected doctor, scientist, professor, or engineer may assume that achievement in one field proves sound judgment in another.

Family members may make the same mistake.

"How can astrology be false if such an educated person believes it?"

The question confuses status with evidence.

Expertise does not transfer automatically.

A cardiologist may know little about astronomy.

A physicist may misunderstand psychology.

A brilliant engineer may have no training in evaluating historical miracle claims.

Intelligence is broad.

Expertise is specific.

An educated person remains human. He still seeks belonging, certainty, hope, and protection from death. He may carry the same childhood fears as anyone else.

Success can even make correction harder.

A respected person becomes accustomed to being trusted. Admitting that a lifelong belief lacks evidence may feel humiliating. If the belief has been defended publicly, the cost of changing becomes greater.

Wisdom begins when achievement no longer protects us from self-examination.

A person may be brilliant in one part of life and frightened in another.

Recognizing that is not an insult.

It is honesty.

Modern Technology, Ancient Fear

Modern technology has not eliminated superstition.

It has given superstition faster transportation.

A family consults an astrologer by video call across continents.

A miracle story reaches millions through social media.

A religious warning is forwarded within minutes.

A confident healer gains authority through polished videos and emotional testimonials.

The device delivering these claims is a triumph of science.

The claims themselves may ignore scientific standards.

A smartphone contains the achievements of physics, engineering, mathematics, and computer science. The same phone may deliver a message warning that bad luck will follow unless it is forwarded to ten people.

Technology changes the speed of transmission.

It does not improve the truth of the claim.

Repeated exposure creates another danger.

A statement seen in many videos begins to feel widely confirmed, even when every video copied the same unsupported source.

Professional lighting creates an impression of expertise.

Confidence replaces qualification.

Testimonials replace complete records.

Algorithms may show people more of what already attracts or frightens them.

The result is an old human vulnerability operating through modern tools.

Access to information is not enough.

People must learn how to judge it.

Who made thc claim?

What evidence is offered?

Can the original source be found?

Does the speaker possess relevant expertise?

Are failed predictions disclosed?

Is fear being used to discourage reflection?

A modern mind is not free merely because it owns modern technology.

Freedom depends on how the information is examined.

What Education Should Produce

A liberating education does more than provide information.

It develops habits of mind.

A genuinely educated person should understand the difference between:

confidence and knowledge,

correlation and causation,

authority and evidence,

possibility and probability,

personal experience and public proof.

Education should also teach consistency.

The standard used to evaluate medicine should apply to faith healing.

The skepticism used toward another religion's miracle should apply to one's own.

The demand for evidence in science should not disappear when a belief becomes emotionally important.

This does not mean that education should impose atheism.

Compulsory disbelief would repeat the error of compulsory faith.

Students should be free to encounter religious and nonreligious ideas, examine their histories, evaluate arguments, and reach their own conclusions.

The purpose is not to dictate the answer.

It is to remove fear from the process.

A mature student should be able to say:

"I may be wrong."

"My tradition may be wrong."

"An authority I respect may be wrong."

"I will change my view if the evidence requires it."

These statements reveal more intellectual freedom than any collection of degrees.

Opening the Door

The divided mind becomes freer when the two rooms are allowed to connect.

A person can begin with one protected belief.

Suppose astrology influences marriage.

He can ask:

What mechanism would allow planetary positions at birth to determine compatibility?

Have astrologers demonstrated reliable predictions under fair conditions?

What happens when astrologers disagree?

Would I accept this method if it came from an unfamiliar religion?

Am I following the chart because it is accurate, or because I fear being blamed?

The same process can be applied to prayer, ritual danger, purity rules, miracle claims, and scripture.

The examination may create discomfort.

An inherited belief carries emotion before it carries argument. Even when the reasoning weakens, the fear may remain.

That is normal.

The goal is not to produce instant certainty in the opposite direction.

It is to remove the wall.

Knowledge from one room is finally allowed to enter the other.

Sometimes the result is rejection of the belief.

Sometimes a ritual remains as cultural symbolism without supernatural claims.

Sometimes the person remains religious but becomes more willing to question authority and abandon harmful practices.

Any of these outcomes may represent progress if they arise from honest examination rather than fear.

A free mind does not have to abandon every tradition.

It must be able to choose.

Education That Reaches the Whole Person

An educated mind remains in chains when knowledge changes a career but never reaches the beliefs a person most wants to protect.

He may operate advanced machines, understand complex biology, and communicate across the world while remaining afraid of a horoscope, ritual omission, curse, or forbidden question.

This contradiction should not be treated with ridicule.

It demonstrates how deeply childhood, identity, and fear shape the mind.

No group escapes these weaknesses completely.

Skeptics can rationalize.

Atheists can become dogmatic.

Scientists can protect favored theories.

Secular ideologies can demand obedience.

The answer is not pride in being more rational than others.

It is a method of correction applied to everyone, including ourselves.

Education becomes liberation when it reaches the sacred room.

It asks the engineer to examine astrology with the seriousness he gives a bridge.

It asks the physician to distinguish spiritual comfort from medical cause.

It asks the professor to read scripture with the same freedom used to examine any other ancient text.

A degree may open a career.

Only self-examination opens the whole mind.

Key Points

- Formal education does not automatically produce intellectual freedom.
- Compartmentalization allows evidence and inherited belief to operate under different rules.
- People may be skeptical professionally while surrendering judgment in emotionally important areas.
- Intelligence can produce sophisticated defenses for beliefs formed before careful examination.

- Family loyalty and fear of exclusion can make honest doubt socially costly.
- Professional status does not provide expertise in evaluating every supernatural claim.
- Modern technology can spread ancient superstition without making it more credible.
- Education becomes liberating when the same standards of evidence are applied consistently.

Reader Reflection

Which beliefs do I examine less carefully because they belong to family or tradition?

Do I apply the same standards at home that I use in my professional life?

What evidence would lead me to change one of my most protected beliefs?

Transition to Chapter 10

A divided mind may remain hidden while belief concerns rituals, stories, or private customs.

The consequences become more urgent when those beliefs enter medicine.

When illness is called punishment, possession, karma, or divine testing, wrong explanations can delay care and deepen suffering.

The next chapter turns to the body, where false belief is no longer only an intellectual problem.

Chapter 10: When Wrong Belief Harms the Body

A man becomes short of breath.

At first, the family hopes it will pass. He is tired, eating less, and struggling to walk across the room. Someone recommends going to the hospital.

Another relative advises waiting.

Perhaps the problem is caused by an unfavorable planet. Perhaps someone cast an evil eye. Perhaps it is karma, divine testing, or a spiritual disturbance. A ritual is arranged. Prayers are offered. A healer is consulted.

The man continues to worsen.

By the time he reaches medical care, his oxygen level is dangerously low. The illness that might have been treated earlier has become more difficult to control.

During decades in medicine, I have seen how quickly fear searches for an explanation. When someone is seriously ill, families often feel helpless. Even educated people may reach first for the beliefs they inherited.

They are not necessarily careless.

They are frightened.

They are trying to do everything possible.

But illness does not pause while a family decides which explanation feels most comforting.

An infection can spread.

A blocked artery can damage the heart.

A stroke can destroy brain tissue.

A psychiatric crisis can become dangerous.

Hours may matter.

Days may become irreversible.

The body does not respond to the sincerity of a belief.

It responds to biology.

Religion becomes dangerous when it moves beyond private comfort and begins directing diagnosis, treatment, or delay.

Prayer may stand beside medical care.

It must not replace it.

The Body Does Not Obey Theology

Religious traditions have offered many explanations for illness.

Disease has been described as punishment for sin, the result of karma, a test from God, possession by spirits, evil eye, curse, or the consequence of ritual impurity.

These explanations arose in societies that did not understand infection, genetics, physiology, neurology, or psychiatry.

They gave frightened people a story when natural causes were unknown.

Historically, that is understandable.

Medically, it is unreliable.

Diabetes develops through biological and environmental factors, not because a deity is displeased.

Cancer does not appear because a person failed spiritually.

Epilepsy is not possession.

Depression is not evidence of moral weakness.

A miscarriage does not reveal divine judgment.

The body follows physical processes whether or not we understand them.

This does not mean medicine knows everything.

It does not.

Diseases remain undiagnosed. Treatments fail. Doctors make mistakes. New evidence changes medical practice.

But uncertainty in medicine does not make a supernatural explanation more accurate.

When a diagnosis is unclear, the honest conclusion is that more evaluation may be needed.

It is not that karma, curse, or divine intention has therefore been established.

A wrong answer does not become helpful merely because the correct answer is incomplete.

Prayer Beside Treatment, or Prayer Instead of Treatment

Prayer can comfort people.

It may calm a frightened patient, bring relatives together, and give someone courage during treatment. Many patients draw emotional strength from faith, and a humane physician should not ridicule that source of support.

The practical boundary is simple:

Is prayer being used beside treatment, or instead of treatment?

A family may pray while calling an ambulance.

A patient may carry religious beads while receiving chemotherapy.

A person may seek spiritual comfort after surgery.

These choices need not conflict with medicine.

The danger begins when prayer is treated as a cure and medical care becomes secondary.

A patient with chest pain waits for a ceremony.

Parents pray over a child with severe fever rather than seeking urgent care.

A person stops medication because a healer declares the illness cured.

A psychiatric emergency is taken to a place of worship instead of a hospital.

In these situations, the problem is not prayer itself.

The problem is the medical claim attached to it.

A comforting practice has been turned into treatment.

Treatment claims require evidence.

If prayer reliably cured infection, cancer, organ failure, or genetic disease, its effects should be visible in medical outcomes.

People of every religion pray with sincerity.

Some recover.

Some do not.

Many receive medical treatment at the same time.

Yet recovery may later be credited mainly to divine intervention, while the unsuccessful prayers disappear from public memory.

Hope is important.

Hope should not be asked to perform the work of medicine.

When Fear Fills the Gap

Families often move toward supernatural explanations when medicine cannot provide immediate certainty.

The test results are incomplete.

The treatment has not worked.

The prognosis is poor.

The physician says, “We do not yet know.”

That answer can feel emotionally unbearable.

A spiritual explanation appears more complete.

"This is God's will."

"Someone has done something to him."

"The illness is karmic."

"The patient is being tested."

The explanation may reduce anxiety because it removes uncertainty.

It can also redirect attention away from what still needs to be done.

Further testing may be delayed.

A second opinion may not be sought.

Symptoms may be interpreted through the spiritual story rather than reported accurately.

The family may spend time and money on rituals while the illness advances.

Medicine sometimes has to live with incomplete answers.

Responsible care does not fill every gap with confidence.

It continues to observe, test, treat, and revise.

The phrase "I do not know yet" may sound weak.

In medicine, it can be a form of honesty that protects the patient.

The Appeal of the Healer

The people most attracted to miracle cures are often those with the most to lose.

They may have advanced cancer, infertility, chronic pain, disability, addiction, or an illness medicine cannot fully cure.

They may be exhausted by treatment.

They may be frightened by side effects.

They may feel that doctors speak in probabilities while the healer speaks with certainty.

Medicine says:

"This treatment may help."

"We cannot promise the outcome."

"There are risks."

The healer says:

"You will recover."

"Your faith will make you well."

"The doctors are wrong."

"The disease is spiritual."

Certainty is emotionally powerful, especially when someone is desperate.

The promise may be accompanied by prayer, touch, sacred water, objects, ceremonies, donations, or instructions to stop treatment.

Some healers may sincerely believe they possess special power.

Others may understand how easily hope can be turned into money and authority.

The patient cannot safely rely on sincerity.

A person claiming the power to cure disease should provide complete records, independent verification, and honest disclosure of failures.

Testimonials are not enough.

Symptoms fluctuate.

Some illnesses enter remission.

A diagnosis may have been wrong.

Treatment already given may begin to work.

Placebo effects can influence pain, anxiety, and subjective symptoms, but they do not reliably eliminate serious disease.

The healer receives credit for visible improvement.

The failed cases disappear.

The danger becomes worse when the patient is blamed for failure.

He did not believe strongly enough.

His faith was weak.

Hidden sin blocked the healing.

The disease remained because his heart was impure.

A patient already carrying illness is made to carry guilt.

No ethical healer should promise what cannot be demonstrated, discourage necessary treatment, or blame the sick for remaining sick.

Compassion without evidence can become exploitation.

Mental Illness Is Not Possession

Few areas reveal the danger of supernatural explanation more clearly than mental health.

A person hears voices.

He becomes suspicious, confused, severely depressed, unusually energetic, withdrawn, or unable to sleep.

His behavior changes in ways that frighten the family.

Where psychiatric knowledge is limited or religious belief is strong, the explanation may be possession, curse, spirit attack, divine punishment, or karma.

The person may be taken to a religious healer instead of a mental-health professional.

Prayer and community support may comfort some patients.

But serious behavioral change requires proper assessment.

Hallucinations, delusions, severe depression, mania, suicidal thinking, and sudden confusion may arise from psychiatric, neurological, medical, or substance-related causes.

Calling the condition possession does not identify the cause.

It may delay treatment.

It may also expose the patient to harm.

People believed to be possessed have been restrained, beaten, isolated, deprived of food, subjected to painful rituals, or prevented from receiving medication.

The family may believe it is fighting an evil force.

The patient experiences fear and injury.

Religious explanation can also deepen stigma.

A health condition becomes a sign of spiritual weakness or moral failure. The person may hide symptoms. Families may avoid seeking help because they fear shame.

This is especially tragic because many psychiatric conditions can improve with appropriate care.

Treatment may involve medication, therapy, social support, attention to physical health, or protection during a crisis.

Psychiatry has limitations.

Medications can cause side effects.

Diagnosis may be difficult.

Those limitations do not make possession a better explanation.

When the mind becomes ill, the person deserves the same dignity given to someone with pneumonia or heart disease.

He needs care.

He does not need an accusation.

Children Cannot Choose the Risk

Adults may choose spiritual treatment for themselves, even when that choice is unwise.

Children are different.

A child depends on adults for protection. He cannot evaluate medical evidence, resist family authority, or obtain treatment independently.

When parents replace necessary care with prayer or ritual, the child carries the consequences of a belief he did not choose.

This creates a difficult boundary between parental freedom and a child's right to health and life.

Parents normally make decisions about food, education, religion, and medical care. That authority is important.

It is not unlimited.

A parent does not own a child's body.

When a serious or treatable condition threatens death or permanent harm, the child's welfare must come before religious doctrine.

Pneumonia, diabetes, meningitis, appendicitis, severe dehydration, and many other conditions can worsen rapidly.

Love does not guarantee a medically sound decision.

Parents who rely on faith healing may be deeply devoted to their children. They may believe that seeking medical care shows spiritual weakness or disobedience.

Their sincerity makes the tragedy more painful.

It does not make the harm less preventable.

Religious freedom should protect belief and peaceful worship.

It should not give adults the right to deny a child effective treatment for a dangerous illness.

A child's survival must not depend on whether the family's theology happens to be correct.

When Sacred Rules Control the Body

Religious beliefs can also influence health through ideas about purity, sexuality, fertility, and duty.

Menstruation may be treated as contamination.

Infertility may be blamed on the woman before either partner receives proper evaluation.

Pregnancy may be surrounded by restrictions that conflict with medical advice.

A person may delay discussing reproductive symptoms because the subject is considered shameful.

The problem is not every cultural practice connected with birth, menstruation, or family life.

Some traditions provide comfort and support.

The harm begins when religious rules create shame, delay care, or take control away from the person whose body is involved.

A girl should not learn that normal biology makes her impure.

A woman should not be prevented from describing pain or bleeding.

A patient should not be forced to risk health in order to satisfy a sacred expectation.

Medical care begins with the person.

Symptoms, informed consent, safety, and dignity matter.

Biology is not impurity.

Pain is not virtue because someone calls it duty.

Blame Becomes a Second Illness

Serious illness often creates a desperate question:

"Why did this happen?"

Religion may answer by assigning moral meaning.

The patient is being tested.

The illness reflects past actions.

The person lacked faith.

The suffering is punishment.

Perhaps the disease will produce spiritual growth.

Some patients may find personal comfort in interpreting illness as a test. That meaning belongs to them.

It should never be imposed on someone else.

From the outside, such explanations may sound hopeful.

To the patient, they may sound like blame.

A person with cancer begins wondering what he did wrong.

Parents of a disabled child carry guilt.

A woman after miscarriage believes she was punished.

Someone with depression concludes that prayer failed because of moral weakness.

These ideas preserve belief in a just universe by making the sufferer responsible for the suffering.

Disease is not distributed according to moral worth.

Kind people develop cancer.

Cruel people remain healthy.

Children are born with genetic disorders.

Devout believers die from infection.

Cells do not examine character before dividing.

Viruses do not ask about faith.

The most humane response is often direct:

"You did not deserve this."

"This is not your fault."

"Let us see what can be done."

These words do not cure the disease.

They remove an unnecessary wound.

Public Health Requires Shared Evidence

Some health decisions affect more than one person.

Vaccination, infection control, sanitation, maternal health, and emergency response depend on collective action.

When religious misinformation enters these areas, private belief can produce public harm.

A person may reject a vaccine because of a rumor or religious warning.

A community may resist treatment because leaders describe it as a threat to faith.

False stories may spread through testimony rather than reliable data.

People have every right to ask questions about medical interventions.

Public-health authorities, physicians, hospitals, and pharmaceutical companies are not beyond criticism.

Medical institutions have made serious mistakes and have sometimes treated vulnerable communities unjustly.

Trust must be earned through transparency, evidence, accountability, and honest communication.

But distrust does not make every alternative claim true.

A frightening story is not stronger than carefully collected evidence.

An illness occurring after vaccination does not automatically prove that the vaccine caused it.

Sequence is not causation.

Public-health decisions should ask:

How common is the risk?

What happens without the intervention?

Are the data complete?

Can the findings be independently checked?

Who is most vulnerable if the disease spreads?

Freedom matters.

So does responsibility for preventable harm.

Why People Turn Away From Medicine

Criticism of faith healing should not turn medicine into another unquestionable authority.

Medicine can be arrogant.

Doctors may communicate poorly, dismiss cultural concerns, overlook symptoms, or treat a person as a diagnosis rather than a human being.

Medical systems may be expensive, rushed, unequal, and difficult to navigate.

Some patients turn to spiritual healers because conventional care left them feeling unheard.

That failure must be taken seriously.

A healer may listen for an hour.

The physician may have ten minutes.

The healer may offer certainty and attention.

The doctor may provide a complicated explanation filled with risk and uncertainty.

The emotional contrast is powerful.

The answer is not to accept unsupported treatment claims.

It is to improve medicine.

Good medical care should combine evidence with listening, humility, and respect.

A physician should explain uncertainty rather than hide behind technical language.

Patients should understand the options, likely benefits, risks, and limits of treatment.

Medicine earns trust by remaining accountable.

A diagnosis can be questioned.

A second opinion can be sought.

A treatment can be revised.

A mistake should be acknowledged.

Neither medicine nor religion should demand blind obedience.

The difference is that medical recommendations must remain open to evidence and correction.

Care Before Creed

When illness enters a family, religious belief may offer emotional support.

Let it support.

It should not diagnose.

It should not replace treatment.

It should not blame the patient.

It should not deny care to a child.

It should not give an unqualified healer authority over a vulnerable body.

The first questions must be practical:

Is the person safe?

Does the condition require urgent evaluation?

What evidence supports the proposed treatment?

Could delay cause harm?

These questions should come before speculation about divine meaning.

A family may pray after calling the doctor.

A harmless ceremony may continue while treatment proceeds.

A patient may draw strength from faith while making medical decisions through informed advice.

The boundary must remain clear.

Spiritual comfort belongs to the individual.

Medical claims belong to evidence.

When the two conflict, care must come first.

I have spent much of my life with patients and families at moments when the body was failing and certainty was impossible.

Medicine could not always cure them.

It could still relieve pain, support breathing, treat infection, explain choices, and protect dignity.

Sometimes the most honest words were:

"We will do everything reasonable."

"I cannot promise the outcome."

"We will not abandon you."

That is not the certainty religion sometimes offers.

It is something more responsible.

Care without false promise.

Key Points

- Illness follows biological processes, not the sincerity of religious explanations.

- Prayer may provide comfort, but it should not replace diagnosis or treatment.
- Faith-healing claims require evidence, especially when patients are told to stop medical care.
- Mental illness should be evaluated medically and psychologically, not treated as possession or spiritual failure.
- Children must be protected when religious decisions place their health or lives at serious risk.
- Religious ideas about purity, fertility, and duty can create shame or interfere with care.
- Patients should not be blamed through claims about sin, karma, weak faith, or divine punishment.
- Medicine must remain humble, accountable, and compassionate while refusing unsupported treatment claims.

Reader Reflection

Have I ever delayed medical care while waiting for prayer, ritual, or spiritual advice?

Do I interpret illness as a moral judgment about the person who is suffering?

When medical advice and religious belief conflict, which one is allowed to guide the body?

Transition to Chapter 11

The consequences of false belief can become severe during illness, but superstition does not appear only in hospitals.

It also enters ordinary decisions about marriage, travel, business, mourning, and family life.

The next chapter examines the quieter ways invisible fear can govern daily choices.

Chapter 11: Invisible Chains: Superstition in Daily Life

The new house is ready.

The documents have been signed. The furniture has arrived. The children have chosen their rooms. Everyone is eager to move.

But the front door remains closed.

An astrologer has said that the day is unfavorable. The family must wait until a particular date and enter at a precise hour.

Work schedules are changed. School plans are rearranged. Relatives are asked to arrive at the approved time. No one wants to take responsibility for ignoring the warning.

Some family members may not fully believe it.

Still, they wait.

"What is the harm?"

"Why upset the elders?"

"Why take a chance?"

This is how superstition often governs modern life.

It does not always arrive through dramatic miracles, possession, or public religious extremism. More often, it appears as a small precaution, a familiar custom, or a warning that seems easier to obey than challenge.

Check the horoscope.

Wait for an auspicious hour.

Do not begin the journey today.

Wear this stone.

Avoid that number.

Perform the ritual so no one can blame us later.

Each decision may appear minor.

Together, they create a way of living in which invisible forces are given authority over visible choices.

The chain is difficult to see because it is made from familiar things: affection, tradition, anxiety, social pressure, and the quiet fear that something may happen if the custom is ignored.

The Power of "What If?"

Many superstitions survive without firm belief.

A person may say that astrology is not scientific and still consult an astrologer before marriage.

A family may doubt that a ritual prevents danger and perform it anyway.

Someone who laughs at lucky objects may carry one during an examination or medical test.

The belief survives through a question:

What if there is something to it?

This sounds cautious.

In reality, it gives an unsupported claim a permanent advantage.

No one can guarantee that misfortune will never follow a rejected ritual. Illness, accidents, conflict, and disappointment occur in every life.

If something unpleasant happens after the warning was ignored, the event may be treated as confirmation.

“We should have listened.”

The countless occasions when nothing happened are forgotten.

Fear has reversed the burden of proof.

Instead of asking whether the superstition is supported by evidence, the doubter is asked to prove that ignoring it will be completely safe.

That proof is impossible.

No decision in life is free from all risk.

The proper question is not whether something bad might happen.

Something eventually will.

The proper question is whether the superstition gives reliable information about that risk.

If an astrologer declares a date dangerous:

Do people who use that date suffer more often?

Do different astrologers agree?

Are failed predictions recorded?

Can the warning be tested fairly?

These questions are rarely answered.

The claim survives because obeying feels emotionally easier than resisting.

Marriage Under the Stars

Marriage is one of the most consequential areas in which superstition can direct family decisions.

Two adults may know each other well. They may share affection, values, goals, and mutual respect. Their families may be supportive.

Then birth charts are compared.

A religious specialist announces that the match is unfavorable.

Suddenly, confidence becomes fear.

One partner may be described as dangerous to the other. The marriage may be predicted to bring illness, conflict, financial loss, or early death.

Remedies may be proposed.

The wedding date must change.

A ritual must be completed.

A stone must be worn.

In some cases, the relationship ends.

The astrologer carries little responsibility for what is lost.

If the couple separates and later regrets it, the chart is not blamed.

If they marry despite the warning and encounter ordinary marital difficulties, the prediction appears confirmed.

If the marriage succeeds, believers may credit the remedy or reinterpret the chart.

The system can explain every outcome.

Meanwhile, the questions that matter most may receive less attention:

Can the couple communicate honestly?

Do they respect one another?

How do they respond to anger and disagreement?

Do they share expectations about children, work, money, religion, and personal freedom?

Is either partner controlling, dishonest, abusive, or irresponsible?

These questions do not guarantee a successful marriage.

They concern the people who will actually live inside it.

Planetary positions do not reveal kindness, maturity, loyalty, or character.

Astrology remains attractive because marriage is uncertain. Families want protection from divorce, conflict, illness, and regret.

A chart gives the impression that hidden risk can be measured before the wedding.

It cannot.

Marriage requires judgment, conversation, consent, and the willingness to grow.

Replacing these with celestial calculations does not remove uncertainty.

It disguises uncertainty as sacred knowledge.

When the Clock Becomes Sacred

Many important events become tied to favorable and unfavorable times.

A business must open at a particular minute.

A journey must begin only after an approved hour.

A family postpones moving, signing a contract, beginning construction, or announcing a pregnancy.

Sometimes the selected time is harmless.

A ceremonial moment may allow relatives to gather and mark an important transition. There is nothing wrong with choosing a meaningful hour when no one is frightened and no practical harm follows.

The problem begins when timing is treated as protection from danger.

A business succeeds through planning, demand, finances, leadership, effort, and chance.

A journey becomes safer through a reliable vehicle, a rested driver, good judgment, and attention to weather and road conditions.

A clock does not protect either one.

The consequences become more serious when astrology enters medicine.

During clinical practice, I have seen families ask whether a procedure could be moved to a more favorable date or hour. They trusted the physician, but they also feared ignoring an astrological recommendation.

They believed they were being careful on two levels.

The two recommendations did not carry equal weight.

One came from the patient's condition, medical urgency, staff availability, and evidence.

The other came from a system that has not demonstrated reliable control over health outcomes.

When there is no medical or practical cost, a family may choose a ceremonial time for cultural reasons.

When delay can affect health, safety, education, employment, or financial stability, evidence must come first.

An hour does not become dangerous because someone declares it inauspicious.

Lucky Objects, Omens, and Dreams

Superstition can attach meaning to ordinary objects and events.

A ring brings success.

A stone protects health.

A number attracts misfortune.

A bird, sneeze, broken object, or person crossing the road becomes an omen.

A dream becomes a warning.

The world begins to feel crowded with hidden messages.

These beliefs gain strength through selective attention.

Once a ring is called lucky, successes associated with it are remembered. Failures are forgotten or explained away.

A disturbing dream followed by bad news feels prophetic. Hundreds of dreams followed by nothing unusual disappear from memory.

The object may also influence behavior.

A student wearing a familiar ring may feel calmer and perform better.

An athlete repeating a routine may gain confidence.

The psychological effect can be real.

The supernatural explanation is unnecessary.

A harmless object may offer comfort without causing difficulty. The problem begins when the person feels unable to act without it.

"I succeeded because I wore the ring."

"I am safe only when I carry this."

"I cannot travel because of the omen."

The object no longer supports confidence.

It replaces it.

Dreams deserve similar caution.

They draw from memory, fear, grief, and unresolved concerns. Because people dream frequently and life contains many events, occasional similarities are inevitable.

The mind connects the dream and the later event after both have occurred.

It rarely records every dream in advance and compares the full record honestly.

A dream may reveal something about our worries.

It does not therefore predict the future.

We can value a personal object without believing it controls reality.

We can reflect on a dream without treating it as supernatural instruction.

Comfort need not become captivity.

Grief, Ritual, and Financial Pressure

Death leaves families vulnerable.

People are grieving, exhausted, and afraid of making a mistake. Religious traditions may prescribe ceremonies, offerings, special foods, repeated observances, pilgrimages, or payments.

Some rituals can help.

They gather relatives.

They provide structure during emotional confusion.

They give people a shared way to remember the dead.

A simple ceremony may offer comfort and dignity.

The burden begins when ritual is driven by fear.

The family may be told that the dead will not find peace unless every requirement is completed.

Failure may bring misfortune to the living.

A ceremony must be repeated.

A particular religious specialist must be paid.

A poor family may borrow money because reducing the ritual feels spiritually dangerous or socially shameful.

Grief becomes an opportunity for pressure.

The people enforcing the custom may be sincere. They may fear the same consequences.

Sincerity does not remove the burden.

The dead do not need the living to become indebted.

A ceremony should serve the grieving.

The grieving should not be forced to serve the ceremony.

Families should be free to choose what they can afford, what reflects the wishes of the deceased, and what offers genuine comfort.

No one should imply that divine mercy depends on payment.

Love is not measured by ritual expense.

A quiet act of remembrance may carry more meaning than an elaborate ceremony organized through fear.

How Family Peace Preserves Superstition

Superstition often survives because challenging it appears more disruptive than obeying it.

A young person disagrees privately but remains silent because an elder will be hurt.

A couple accepts an unnecessary ritual to prevent conflict before a wedding.

A daughter follows an exclusion rule because refusing would upset her mother or grandmother.

The immediate calculation seems reasonable:

“It is only one ceremony.”

“Let us keep everyone happy.”

"There is no point arguing."

Compromise is part of family life.

Not every symbolic custom deserves a confrontation. People may participate out of affection rather than belief.

That can be a reasonable choice.

But repeated silence has consequences.

Children watch.

They see adults become anxious about dates, omens, planets, impurity, and ritual mistakes. Even when parents say, "We do not really believe this," their behavior suggests that the danger may be real.

Fear is transmitted through tone, hesitation, and urgency.

The child learns which customs cannot safely be refused.

Later, he may repeat them for the same reason: not conviction, but family peace.

This is how a superstition can survive even when few people fully believe it.

Each person assumes that someone else believes more strongly.

Breaking the pattern requires honesty without unnecessary cruelty.

A family member can say:

"We respect the people involved, but we do not accept this prediction."

“We will join the celebration, but we will not frighten the children.”

“We will not delay medical care.”

“We prefer a simple ceremony that we can afford.”

Love does not require pretending that an unsupported claim is true.

Sometimes family peace becomes healthier after the first discomfort passes.

Superstition on a Screen

Modern technology has not eliminated superstition.

It has accelerated it.

A warning about a dangerous day reaches thousands of people within hours.

A miracle video circulates without context.

An astrologer offers consultations across continents.

A message promises blessing if forwarded and misfortune if ignored.

The old fear appears on a new screen.

Digital superstition often uses urgency:

“Do this tonight.”

“Do not ignore this warning.”

“Share immediately to protect your family.”

The message discourages reflection by making delay feel dangerous.

Repetition creates another illusion.

When the same claim appears in several videos or messages, it may seem independently confirmed. In reality, every source may have copied the same rumor.

Professional editing can also create false authority.

A confident speaker with polished graphics and thousands of followers may appear knowledgeable even when no evidence is offered.

Education does not guarantee protection.

Doctors, engineers, professors, and other professionals also forward fear-based messages when the subject involves children, health, marriage, money, or death.

Before sharing, pause.

Who produced the claim?

Can the original source be identified?

What evidence is offered?

Is the message using fear to force immediate action?

Would I believe it if it came from another religion or culture?

A warning that depends on panic often weakens when given time.

One of the most useful responses to digital superstition is simple:

Do not forward it.

Silence can interrupt the chain.

A Practical Test for Superstition

Not every custom requires scientific testing.

A family meal, song, symbolic blessing, or holiday ceremony may be valuable because of what it means to people.

Testing becomes necessary when a practice claims power over events or places pressure on others.

Ask:

What is supposed to happen?

How would we know whether the practice worked?

What happens when the prediction fails?

Are unsuccessful cases remembered?

Would we accept the same reasoning from another tradition?

Is anyone being frightened, excluded, delayed, or made to pay?

What would we choose if no one were afraid?

That final question is often the most revealing.

A family may discover that it would keep the ceremony because it is beautiful and meaningful.

Then the tradition can remain without supernatural threat.

It may discover that no one values the ritual except as insurance against invisible danger.

Then fear has become the real reason.

The purpose of questioning is not to empty life of ceremony.

It is to make participation voluntary.

A freely chosen tradition can bring joy.

A tradition obeyed because refusal feels dangerous is a chain.

Living Without Invisible Permission

Superstition teaches people to seek permission from forces that cannot be shown to exist.

The stars must approve the marriage.

The date must approve the journey.

The omen must approve the decision.

The ritual must approve the future.

A freer life returns responsibility to human beings.

We choose a partner by examining character and compatibility.

We open a business after careful planning.

We schedule treatment according to medical need.

We remember the dead through love rather than fear.

We accept that no ring, date, chart, or ceremony can remove every uncertainty from life.

This freedom may feel uncomfortable at first.

A person raised with superstition may ignore an astrological warning and remain anxious throughout the day. If something unpleasant happens, the old belief may return immediately.

That reaction does not prove the warning.

It shows how deeply the association was learned.

The chain weakens through experience.

A journey begins at an "unfavorable" hour and ends normally.

A marriage proceeds despite a chart and grows through mutual respect.

A family simplifies a ceremony, and no supernatural punishment follows.

Life gradually reveals what fear concealed.

The superstition was not protecting us.

It was teaching us to ask permission.

Key Points

- Superstition often survives through the question, "What if something happens?"
- Astrology can displace meaningful evaluation of character, consent, and compatibility in marriage.
- Auspicious timing becomes harmful when it delays medical, financial, or practical decisions.
- Lucky objects, omens, and dreams gain credibility through selective memory and emotional association.
- Ritual can support grief, but fear-based ceremonies may burden families emotionally and financially.

- Family peace can preserve beliefs that few people privately accept.
- Digital technology spreads superstition rapidly through repetition, urgency, and fear.
- A healthy tradition remains meaningful when participation is voluntary and supernatural threats are removed.

Reader Reflection

Which ordinary decision in my family is influenced by astrology, omens, or ritual fear?

Would we continue a particular custom if no one feared the consequences of stopping it?

What belief are children learning from our behavior, even when we say we do not accept it?

Transition to Chapter 12

Daily superstition may appear harmless when attention remains on the ritual.

Its human cost becomes clearer when we look at the people expected to obey, pay, remain silent, or carry the blame.

The next chapter turns from the belief to the person who bears its consequences.

Chapter 12: The Victims of Sacred Fear

A girl has her first menstrual period.

A few days later, the family gathers for a religious ceremony. Lamps are lit. Flowers are arranged. Prayers begin.

The girl is told to remain outside the prayer room.

No one shouts at her. The instruction may come gently from a mother or grandmother who lived under the same rule.

"It is only for a few days."

"It is our tradition."

"Do not take it personally."

But the girl understands more than the words say.

Something about her body has made her unfit to stand beside the others.

She watches from the doorway while the family continues without her.

This is how sacred fear often enters a life.

It does not always arrive through open cruelty. Sometimes it appears as a familiar rule spoken by people who genuinely love the person they are excluding.

That combination makes the injury harder to name.

Affection and humiliation can exist in the same room.

Religion speaks of purity, duty, karma, protection, obedience, and family honor. The person living under the rule may experience something very different:

shame,

fear,

silence,

exclusion,

debt,

or blame.

To understand the human cost of religious belief, we must look beyond the doctrine and ask a more direct question:

Who is required to carry it?

Fear Falls Downward

Sacred fear is rarely distributed equally.

Its heaviest burden usually falls on people with the least authority: children, women, widows, lower castes, the poor, the sick, doubters, minorities, and social outsiders.

The powerful interpret the rule.

The vulnerable live under it.

A religious leader declares that a ceremony is necessary.

A poor family must find the money.

An elder defends purity.

A girl must accept exclusion.

A community fears witchcraft.

An isolated woman becomes the accused.

A family insists that belief creates unity.

The doubter must hide.

People with status can often escape the harshest consequences of the beliefs they defend. They possess money, education, family protection, or the authority to reinterpret a rule when it becomes inconvenient.

Those with less power receive fewer exceptions.

A wealthy family may perform expensive rituals without sacrificing medicine or food. A poor family may borrow money.

A respected man may quietly ignore a religious restriction. A young woman may be judged publicly for doing the same.

A powerful spiritual leader may claim special knowledge. A frightened patient who questions him may be accused of lacking faith.

Sacred language can make this inequality appear moral.

Obedience becomes virtue.

Resistance becomes selfishness.

Exclusion becomes purity.

Submission becomes duty.

Whenever a religious rule falls mainly on those least able to resist, we should ask whether the rule is protecting the sacred or protecting power.

Children Who Learn to Fear Themselves

Children trust adults before they can evaluate what adults teach.

A young child cannot examine the evidence for hell, karma, possession, divine surveillance, or supernatural punishment. These ideas come from parents, teachers, relatives, and religious authorities, the same people who provide affection, protection, and approval.

To the child, love and truth are closely connected.

"God is always watching."

"Bad thoughts are sinful."

"Doubt may lead you away from God."

"People who do not believe will be punished."

Adults may intend to teach morality.

The child may learn something more troubling:

parts of his own mind are dangerous.

Curiosity becomes disobedience.

Anger becomes sin.

Sexual development becomes shame.

Doubt becomes guilt.

The child begins policing himself long before an authority needs to be present. He fears punishment for thoughts he did not choose and feelings he does not yet understand.

These fears may survive after belief has weakened.

An adult may no longer accept hell intellectually and still feel panic when thinking about death. He may reject a ritual but experience guilt when refusing it. He may hide disbelief because honesty feels like betrayal of his parents.

The educated adult continues carrying the frightened child.

Children need moral guidance. They should learn honesty, empathy, fairness, responsibility, and concern for others.

But morality should grow from understanding harm and respecting people.

It should not depend on terror of invisible punishment.

A child's conscience should be developed.

His imagination should not be terrorized.

Women Made Responsible for Purity

Religious communities have often placed the burden of purity and family honor on women.

The rules vary across traditions and cultures, but familiar patterns appear around menstruation, clothing, sexuality, fertility, marriage, obedience, and widowhood.

The woman is expected to adjust.

A menstruating girl is excluded.

A wife is told that family peace depends on her submission.

A woman without children is pitied, blamed, or taken from one religious remedy to another.

A widow may be treated as unlucky or excluded from celebrations.

These practices may be defended as respect, protection, modesty, or complementary duty.

The test is not the beauty of the explanation.

The test is the life the woman is permitted to live.

Can she refuse without punishment?

Does the rule apply equally to men?

Is her consent meaningful?

Does the practice protect her dignity and health?

A restriction does not become equal because someone calls male and female roles different but equally valuable.

If one person decides and the other must obey, the moral imbalance remains.

Men are rarely treated as spiritually contaminating because of normal biology. Widowers are seldom removed from celebration as symbols of misfortune. Male sexuality is rarely governed with the same intensity or consequence.

The inequality reveals what the language of purity conceals.

A woman's body is not a spiritual danger.

Menstruation is not contamination.

Infertility is not punishment.

Widowhood is not bad luck.

No woman should be required to apologize for being fully human.

When Birth Becomes Destiny

Caste shows how religious ideas can give human inequality the appearance of sacred order.

Caste has a complex history. Religion is not its only cause. Political power, occupation, property, family structure, and local custom have also sustained it.

But ideas about karma, inherited duty, purity, and pollution have repeatedly been used to justify hierarchy.

A person may be taught that social position reflects actions from a previous life.

Certain duties belong to certain births.

Contact with another group creates impurity.

A human arrangement begins to look cosmic and inevitable.

The consequences are not theoretical.

Birth can influence marriage, occupation, social acceptance, access to opportunity, entry into religious spaces, and the dignity granted by others.

A child may learn early that some people are naturally higher and others naturally lower.

Sacred explanation makes resistance more difficult.

If inequality was created by people, it can be changed by people.

If it is described as karma or divine order, resistance appears to challenge the universe itself.

The oppressed person must then fight not only social power, but the claim that humiliation carries spiritual meaning.

No child is born impure.

No human touch is contaminating.

No occupation reduces a person's worth.

No sacred story should turn the accident of birth into a lifelong sentence.

The Poor, the Sick, and the Accused

Sacred fear becomes especially cruel when it adds a second burden to someone already suffering.

A poor family is not only grieving. It is told that an expensive ceremony is necessary for the peace of the dead.

A patient is not only facing illness. He is asked whether sin, karma, or weak faith caused it.

A person with mental illness is not only confused or frightened. He is called possessed.

A widow is not only mourning. Her presence is treated as unlucky.

An isolated woman is not only vulnerable. She becomes the suspected source of a curse.

The explanation may comfort the wider community because it gives suffering a cause.

But the cause is often placed on the person least able to defend himself.

During my years in medicine, I have seen patients carry guilt that illness did not create. They wondered whether their disease was punishment, whether they had failed spiritually, or whether the family's prayers remained unanswered because of them.

Disease was already enough.

The moral accusation added pain no patient should have been asked to bear.

The same process appears when a frightened community searches for a scapegoat.

A child dies.

Crops fail.

Disease spreads.

Several misfortunes occur close together.

Instead of accepting uncertainty or investigating natural causes, suspicion gathers around someone who is poor, widowed, elderly, mentally ill, isolated, or socially different.

Once the person is treated as the source of danger, ordinary compassion weakens.

Fear sees a threat where humanity should see a person.

The Doubter Who Must Hide

Religious unity often looks peaceful because disagreement has been pushed underground.

A family may proudly describe itself as devout while one member remains silent about disbelief.

He attends ceremonies.

He repeats familiar words.

He avoids questions.

He allows others to assume that he believes.

The outward harmony may depend on his concealment.

Honesty could disappoint parents, create conflict with a spouse, affect marriage prospects, or lead to exclusion from family and community life.

The person is told that religion holds the family together.

What this may mean in practice is that one person must hide in order to preserve everyone else's comfort.

Belonging becomes conditional.

You may remain fully loved if you believe correctly.

You may be tolerated if you remain silent.

You may be rejected if you speak honestly.

This pressure is especially powerful because it is rarely described as coercion.

No one may threaten physical punishment. The family may simply withdraw warmth, express grief, or treat disbelief as personal betrayal.

The doubter carries guilt for causing pain by telling the truth.

A belief that requires emotional punishment to keep people inside it should not be confused with freely chosen faith.

Love should be strong enough to survive disagreement.

A family has not protected unity if it has made honesty unsafe.

Harm Is Often Passed Down by Loving People

It would be easier if every harmful religious practice were imposed by obviously cruel people.

Often it is not.

A mother teaches her daughter an impurity rule because her own mother taught it to her.

A father frightens a child with hell because he believes fear will protect the child's soul.

A grieving relative insists on an expensive ceremony because he fears what may happen if it is omitted.

A community excludes a doubter because leaders believe unity depends on shared faith.

The people transmitting the harm may sincerely believe they are protecting those they love.

This does not make them evil.

It does not make the practice harmless.

Good intentions can carry harmful ideas from one generation to the next.

Understanding this allows us to challenge the belief without dehumanizing the believer.

The mother may also be a victim of the rule she now enforces.

The father may still carry fears planted in his own childhood.

The elder may never have been allowed to question the tradition.

Compassion should extend in both directions.

We can understand why people repeat a harmful belief.

We should not continue the harm merely because its carriers were sincere.

An inheritance can be understood without being preserved.

The Person Before the Belief

Religious discussions often remain abstract.

People debate scripture, tradition, purity, morality, and divine will.

The person carrying the consequence disappears.

Bring that person back into view.

Look at the girl standing outside the prayer room.

Look at the child frightened by hell.

Look at the patient blamed for illness.

Look at the poor family borrowing money for a ritual.

Look at the widow treated as unlucky.

Look at the doubter pretending in order to remain loved.

Then ask:

What is this belief doing to this person?

Does it protect dignity or reduce it?

Does it offer freely chosen meaning or demand submission?

Can the person refuse without fear?

Who gains authority?

Who pays the price?

These questions do not settle every religious argument.

They expose the moral consequence.

A tradition does not become humane because it is ancient.

A restriction does not become respectful because it is spoken gently.

A fear does not become true because it was inherited.

A sacred explanation does not become just because the victim has been taught to accept it.

When belief and human dignity collide, the human being must come first.

What Ends With Us

Large systems often change through small acts of refusal.

A parent allows the difficult question.

A family says a menstruating girl is not impure.

A patient seeks treatment instead of accepting blame.

A couple rejects an astrological warning.

A poor household chooses a simple ceremony it can afford.

A believer tells a doubting child, “You will still be loved.”

These acts may appear small.

Each interrupts an inheritance.

Change often begins when someone says:

“She is not impure.”

“We will not blame him.”

“Our child may ask questions.”

“We will not be frightened into paying.”

“I love you, but I do not believe this.”

The first refusal may create tension.

Elders may feel hurt.

Relatives may predict danger.

The person breaking the custom may feel guilty even when he knows the rule is unjustified.

That discomfort is part of liberation.

It does not prove the tradition was true.

It shows how closely the tradition had been tied to fear and belonging.

The purpose of reason is not merely to win arguments about God or scripture.

It is to reduce avoidable suffering.

Reason protects people from shame, accusation, exploitation, and inherited control.

Once the person harmed becomes visible, neutrality becomes harder.

We must decide whether sacred beliefs exist to serve human beings or human beings exist to serve sacred beliefs.

A humane society places the living person first.

Key Points

- Sacred fear usually places its greatest burdens on people with the least social power.
- Children may learn to fear their own thoughts, bodies, and questions.

- Religious ideas of purity and honor often impose unequal restrictions on women.
- Sacred explanations can turn inherited social inequality into apparent moral or cosmic order.
- The poor, sick, widowed, mentally ill, and socially isolated may be blamed or exploited when suffering receives supernatural meaning.
- Doubters lose freedom when family love and belonging depend on approved belief.
- Harmful customs are often transmitted by loving people who believe they are protecting others.
- When doctrine and human dignity conflict, the human being must come first.

Reader Reflection

Who carries the greatest burden of the religious customs in my family or community?

Have I ever defended a tradition more carefully than the person harmed by it?

Can someone in my family reject belief without losing love and belonging?

Which sacred fear should end with my generation?

Transition to Chapter 13

This book has followed religious fear from the first forbidden question to the people who carry its consequences.

The final chapter turns toward what becomes possible when fear loses its authority: a mind that can live with uncertainty, question without guilt, and choose human dignity without asking sacred permission.

Chapter 13: A Mind No Longer Afraid

A child asks his father, “What if God is not real?”

The father feels the old answer rising inside him.

Do not say that.

God may hear you.

Good children should not question such things.

Those words came to him in childhood. For years, they lived in his mind so quietly that they felt like his own.

He looks at his child and pauses.

Then he says, “That is an important question. We should be able to ask it.”

Nothing dramatic happens.

No ritual stops. No relative changes his mind. No religious institution loses its power that afternoon.

But something has changed.

A fear has reached the next generation and has not been passed forward.

Sometimes freedom begins that quietly.

The First Freedom Is Honesty

Religious belief is rarely only an opinion about God.

It becomes attached to childhood, family, language, festivals, food, music, marriage, grief, and memories of people we love. Questioning religion can therefore feel like questioning an entire life.

That is why many people continue outwardly long after belief has weakened inwardly.

They repeat prayers.

They attend ceremonies.

They avoid difficult conversations.

They tell themselves they are merely preserving family peace.

Sometimes they are. Not every disagreement requires confrontation. A person may participate in a harmless celebration out of affection rather than belief.

But the private mind knows when it is pretending.

The first freedom is not a public declaration.

It is honesty with oneself.

A person may finally admit:

“I do not find this convincing.”

“I participate because my family expects it.”

"I am afraid of what may happen if I refuse."

"I have defended this belief because it belongs to my identity."

Such admissions can bring relief and grief together.

A familiar certainty is being lost. So is the feeling of belonging completely within the religious world of childhood.

I did not reach my own conclusions in one dramatic moment. I inherited religious ideas before I could examine them, as most people do. Questioning them required more than learning scientific facts.

It required separating affection for people from acceptance of their beliefs.

That distinction was not always easy.

We can love our parents without believing everything they taught us.

We can remember our grandparents warmly without repeating every fear they inherited.

We can value a festival, a language, a song, or a family memory without granting supernatural authority to the beliefs surrounding it.

Intellectual honesty does not require emotional rejection.

It requires the courage to stop calling something true merely because it is familiar.

Doubt Is a Safeguard

Religion often presents doubt as temptation, weakness, rebellion, or moral failure.

A freer mind understands doubt differently.

Doubt is not the refusal to believe anything.

It is the willingness to admit that we may be mistaken.

That willingness protects us.

It slows us before we give money to someone selling miracles.

It asks for medical evaluation before accepting possession or curse.

It notices when an astrologer speaks vaguely, when a failed prediction is reinterpreted, and when a literal religious claim becomes symbolic only after evidence contradicts it.

In medicine, doubt can prevent a premature diagnosis.

A physician may recognize a familiar pattern and still ask whether something does not fit. A different disease may produce similar symptoms. A test may have been misleading. New evidence may require a change in treatment.

Certainty feels efficient.

Correction protects the patient.

The same discipline should apply to religious claims, especially when they involve invisible powers, divine authority, eternal punishment, money, health, or control over another person.

A mind no longer afraid does not say, "I will trust nothing."

It says:

"I will not surrender my judgment merely because uncertainty makes me uncomfortable."

Doubt should not become another rigid identity. Skeptics can also become arrogant, selective, or unwilling to reconsider.

The purpose is not permanent suspicion.

It is proportion.

Confidence should grow with evidence and weaken when evidence fails.

Meaning Without Sacred Permission

Many people fear that losing religious certainty will make life empty.

If there is no guaranteed divine plan, why are we here?

If death may be final, what makes life valuable?

If morality does not come from God, why should anyone be good?

These questions deserve serious answers.

Religion has supplied language for meaning, community, grief, and morality for generations. When belief weakens, a person may initially feel as though the structure holding life together has collapsed.

But meaning does not belong exclusively to religion.

People find purpose in love, family, friendship, work, creativity, discovery, service, and the effort to reduce suffering.

A physician does not need a divine command to recognize that pain matters.

A parent does not need scripture to love a child.

A friend does not need a promise of heaven to sit beside someone who is dying.

Morality grows from human capacities we can recognize directly: empathy, cooperation, experience, reason, and the awareness that other people can suffer as deeply as we do.

Religion may express these capacities.

It did not create or own them.

A finite life can be deeply meaningful precisely because it is finite.

Time with people we love matters because it cannot be repeated forever.

Relieving suffering matters because no future world can be relied upon to repair what we neglect in this one.

Justice becomes more urgent when we stop assuming that another universe will correct our failures.

A life without sacred certainty does not lose love, beauty, gratitude, music, courage, or responsibility.

It loses only the claim that these human realities require supernatural permission.

Freedom Does Not Require the Destruction of Culture

Leaving unsupported religious belief does not require erasing every tradition.

Culture is larger than theology.

Families may preserve food, music, language, stories, art, celebration, and remembrance. A ceremony may continue as an expression of belonging rather than fear.

The difference lies in freedom.

Can people participate without being threatened?

Can they decline without guilt?

Can a child ask what the ritual means?

Can a woman refuse a rule that humiliates her?

Can a poor family choose a simpler ceremony without being told that the dead will suffer?

A tradition freely chosen may enrich life.

A tradition enforced through supernatural fear becomes control.

We do not have to destroy the lamp because we reject the claim that failing to light it will bring misfortune.

We do not have to reject our families because we reject an inherited superstition.

We can preserve memory while allowing truth to change.

That is not cultural betrayal.

It is cultural maturity.

Criticism Without Cruelty

Freedom from religion should not become contempt for religious people.

Many believers are kind, thoughtful, and morally serious. Many inherited their beliefs before they were old enough to choose. Some remain religious because faith gives them comfort during illness, grief, poverty, or loneliness.

Others have challenged harmful traditions from within their faiths and helped their communities become more humane.

Recognizing this does not weaken criticism.

It makes criticism more honest.

A humane critic can say:

"I respect you, but I do not accept this claim."

"I understand why this comforts you, but comfort does not establish truth."

"I value our family, but I will not teach this fear to my child."

"We can preserve the celebration without preserving the superstition."

Firmness and compassion can exist together.

Believers should not be humiliated.

Harmful beliefs must still be challenged.

The purpose of reason is not to create a new tribe of people proud of their superiority.

It is to help us correct error and reduce suffering.

Freedom Must Be Practiced

Religious fear rarely disappears the moment belief changes.

A person may reject astrology and still feel uneasy after ignoring a warning.

He may stop believing in hell yet continue to feel childhood anxiety about death.

He may understand that a ritual has no demonstrated power and still fear that something terrible will happen if he refuses it.

This does not prove that the belief remains true.

It shows that emotion changes more slowly than thought.

Ideas learned early live in memory, habit, family relationships, and bodily reactions. Freedom therefore develops through practice.

Ignore one unsupported warning.

Let a coincidence remain a coincidence.

Refuse to forward one fear-based message.

Choose medical care without delay.

Allow a child to question.

Reject one humiliating purity rule.

Speak one honest sentence.

Each act teaches the mind that reality does not collapse when superstition is disobeyed.

At first, the refusal may feel reckless.

Later, it feels ordinary.

Eventually, the person recognizes that the fear was never protecting him.

Obedience merely prevented him from discovering that he was already free.

The Strength of Not Knowing

A mind no longer afraid does not need a final answer for every mystery.

It can face uncertainty without immediately filling it with God, destiny, karma, spirits, or divine purpose.

It can admit that questions about death, consciousness, and the ultimate origin of the universe remain difficult.

This is not intellectual emptiness.

It is honesty about the limits of knowledge.

Religion often explains suffering too quickly. It tells grieving people why someone died, why a child became ill, or what lesson a tragedy was meant to teach.

A freer mind can resist that impulse.

It can say:

"This is terrible."

"I do not know why it happened."

"You did not deserve it."

"Let us help."

These sentences offer less cosmic certainty.

They offer more human presence.

During my medical career, I often had to speak when no perfect answer was available. Treatment could not always cure. A diagnosis was not always immediate. Families understandably wanted assurance that no honest physician could give.

Uncertainty did not require abandonment.

We could still relieve pain, explain what was known, discuss the choices, and remain beside the patient.

"I do not know" and "I will not leave you" can exist in the same conversation.

That combination is more compassionate than false certainty.

We do not need to understand the universe completely before caring for one another within it.

Responsibility Returns to Us

Religion often promises that a higher power governs the final outcome.

Justice will be completed.

Suffering will receive meaning.

The good will be rewarded.

The dead will live again.

A secular understanding offers no such guarantee.

That can be frightening.

It also places responsibility where it belongs.

If children are to be protected, human beings must protect them.

If patients are to receive care, human beings must provide it.

If caste, gender inequality, persecution, and superstition are to end, human beings must challenge them.

We cannot postpone justice to another world.

We cannot excuse suffering as part of a divine plan.

We cannot wait for prayer to perform the work that requires human action.

The absence of a guaranteed cosmic rescue does not make life hopeless.

It makes our choices more important.

No god needs to command compassion before suffering deserves a response.

Pain is already real.

That is enough.

What We Owe the Next Generation

A child should not be forced to choose between intellectual honesty and family love.

He should be able to ask:

"How do we know?"

"Why do other religions believe something different?"

"What evidence would show that this claim is wrong?"

"Why must I obey this ritual?"

The adult does not need to have every answer.

He can say:

"I do not know."

"People believe different things."

"Let us examine the evidence."

"You will still be loved whatever conclusion you reach."

A child should learn that authority can be mistaken, that confidence is not proof, and that ancient ideas may be questioned.

He should also learn empathy.

Reason without concern for people can become cold and self-satisfied. The purpose of questioning is not merely to expose error.

It is to protect human beings from the consequences of error.

A child allowed to think is not being deprived of morality.

He is being given the tools to develop a conscience that belongs to him.

A Mind No Longer Afraid

This book began with a child's question:

How do we know this is true?

That question led us through the psychology of belief, the attraction of certainty, the transformation of fear into sacred duty, the weakness of miracle evidence, the special exemption granted to religion, the authority of scripture, and the correction provided by science.

It also led into ordinary human life:

the educated person who keeps reason in separate rooms,

the patient endangered by false belief,

the family directed by superstition,

and the vulnerable person required to carry the burden of sacred fear.

The conclusion is not that every believer is foolish or cruel.

Many are sincere and compassionate. Many are themselves carrying beliefs planted before they could choose. Some have resisted harmful authority and reformed traditions from within.

The central problem begins when a belief is protected from correction.

Religion becomes the enemy of reason when it teaches people to stop asking, gives fear the authority of truth, places scripture above evidence, or treats doubt as moral failure.

A mind no longer afraid reverses that order.

It permits the question.

It follows the evidence.

It changes when correction is needed.

It refuses to defend a doctrine more carefully than the human being living under it.

This freedom is not perfect knowledge.

It is not a new sacred certainty.

It does not end grief, error, conflict, or death.

It offers something more modest and more dependable:

the freedom to think honestly,

the courage to live without invented answers,

and the responsibility to place human dignity above sacred fear.

The child asks, “What if God is not real?”

A mind no longer afraid does not punish the question.

It opens the door.

Key Points

- Intellectual freedom often begins with private honesty about what we truly believe.
- Doubt protects us by keeping important claims open to correction.
- Meaning, morality, culture, and love do not depend on supernatural certainty.
- Believers should be treated humanely while harmful beliefs are challenged clearly.
- Religious fear weakens through repeated acts of honest refusal.
- “I do not know” can express courage, humility, and compassion.

- A life without belief in divine control places greater responsibility on human beings.
- Children should receive the freedom and skills to examine religious claims for themselves.

Reader Reflection

Which religious fear still carries authority inside my mind?

What belief am I protecting because it is connected to family, identity, or comfort?

Can I criticize religion firmly without losing compassion for believers?

What fear will I refuse to pass to the next generation?

Transition to Book 2

This book has examined how religion enters and governs the individual mind.

But no mind develops alone.

Families teach belief. Communities reward it. Rituals repeat it. Institutions protect it. Entire societies can treat inherited claims as obvious truth long before a child is capable of examining them.

Book 2, Religion: The Mass Delusion, turns from the individual mind to the collective one. It examines how societies teach, normalize, defend, and pass down belief without evidence.

Appendix A: Common Religious Claims and Clear Answers

Most objections to this book will not arrive as formal theology.

They will come through familiar sentences heard in homes, schools, hospitals, places of worship, and moments of fear:

“Religion gives people hope.”

“Science does not know everything.”

“Faith is another way of knowing.”

“Without religion, people would become immoral.”

“What harm is there if belief brings comfort?”

Some of these statements contain partial truths.

Religion can comfort people. Science does not know everything. Religious communities have produced art, charity, courage, and social belonging. Rituals can bring families together. Faith may help a person endure grief or uncertainty.

The problem begins when a partial truth is used to protect a larger claim from examination.

The answers below are intentionally brief. The fuller arguments have been developed throughout this book.

1. "Religion comforts people. Why take that away?"

Comfort matters.

A grieving person may find strength in prayer. A frightened patient may feel less alone. A religious community may offer companionship, practical help, and familiar language during loss.

None of this should be mocked.

But comfort and truth are different questions.

A belief can reduce fear without accurately explaining reality. A child may feel safer after being told that a parent who died is watching from heaven. The comfort may be real. The claim still requires evidence.

Religious comfort also carries different consequences for different people. One believer finds peace in prayer. Another is terrified by hell, blamed through karma, or pressured into an expensive ritual.

The goal is not to remove comfort unnecessarily.

It is to separate comfort from unsupported claims, fear, blame, and control.

People deserve consolation without deception.

2. "Science does not know everything."

That is true.

Science does not fully explain consciousness, the earliest conditions of the universe, or every feature of life and human experience. Scientific conclusions remain incomplete and open to correction.

But an unanswered question does not establish a religious answer.

“We do not know” and “therefore God did it” are not equivalent.

The first acknowledges a limit.

The second fills the limit with an unsupported conclusion.

History repeatedly shows the danger of doing this. Disease, lightning, planetary movement, and mental illness were once explained through supernatural forces. Natural investigation later produced better explanations.

Science does not become weak by admitting uncertainty.

It becomes more trustworthy.

An honest gap in knowledge is better than a confident invention.

3. “Faith is another way of knowing.”

Faith may be a way of trusting, hoping, or committing oneself to a religious tradition.

It is not a reliable method for determining which conflicting religious claim is true.

Christians, Muslims, Hindus, Sikhs, and followers of other religions may all appeal to faith while reaching incompatible conclusions.

Their sincerity may be equal.

Their conclusions cannot all be correct in the same way.

If faith can support opposite answers, faith cannot tell us which answer corresponds to reality.

Faith may explain why someone believes.

It does not independently establish that the belief is true.

4. "I have experienced God personally."

A personal experience can be profound.

A person may feel a presence during prayer, receive comfort after asking for guidance, or experience unity during meditation. The event may change his life.

The experience should not be dismissed merely because it is religious.

But the experience and its interpretation are not identical.

People from different religions interpret similar experiences through different beliefs. A Christian may experience Jesus. A Hindu may experience Krishna. Another person may interpret the same sensation through grief, expectation, meditation, or psychology.

A person can honestly say:

"I felt something powerful."

"It gave me strength."

"It changed me."

The experience alone cannot establish:

"My explanation is therefore true for everyone."

Private meaning deserves respect.

Public claims require evidence others can examine.

5. "Miracles prove that God exists."

An unusual event is not automatically a miracle.

A patient may recover unexpectedly. A prediction may appear accurate. A statue may seem to behave strangely. A prayer may be followed by good news.

Before calling the event supernatural, we must ask whether natural recovery, coincidence, misunderstanding, incomplete records, selective memory, expectation, or fraud could explain it.

We must also count failures.

If one recovery after prayer is evidence for divine intervention, what do we do with the many equally sincere prayers followed by death?

Every major religion reports miracles, often using them to support contradictory doctrines.

A fair question is:

Would I accept this evidence if the miracle supported another religion?

An unexplained event remains unexplained.

It does not automatically prove one community's theology.

6. "Scripture contains great wisdom. Does that not prove it came from God?"

Sacred texts contain poetry, moral insight, philosophy, history, memory, and profound observations about human life.

So do many works of human literature.

Wisdom does not prove divine authorship.

A book may contain compassionate teachings alongside scientific errors, historical limitations, harsh commands, or social assumptions from the age in which it emerged.

Modern believers already use moral judgment when reading scripture. They preserve some passages, reinterpret others, and reject many in practice.

This shows that the reader brings conscience to the book.

A wise teaching should be accepted because it is wise.

A harmful command should be rejected because it is harmful.

Neither conclusion requires the book to be perfect.

7. "Science and religion answer different questions."

Sometimes they do.

Science examines the natural world. Religion may offer ceremony, identity, personal meaning, or symbolic language.

Conflict is limited when religion remains in those areas.

But religion often makes factual claims:

God created life in a particular way.

Prayer changes physical outcomes.

Miracles suspend natural law.

Illness reflects punishment or possession.

Scripture accurately describes history or nature.

Once religion makes claims about the observable world, those claims enter the territory of evidence.

Religion cannot make factual statements and then retreat to "different questions" only when the evidence becomes difficult.

Symbolic meaning may coexist with science.

Claims about reality must answer to reality.

8. "Without religion, people would become immoral."

Religious people can be deeply moral.

That does not show that morality depends on religion.

Human beings can recognize suffering, fairness, loyalty, care, and injustice through empathy, reason, experience, and social life.

Parents do not need a divine command to love their children.

Physicians do not require belief in hell to care for patients.

Friends do not need a promise of heaven to remain beside someone who is dying.

Religious scriptures also contain conflicting moral rules. Believers already decide which teachings to follow by using judgment shaped by values beyond scripture.

Morality should be judged by what actions do to conscious beings.

Cruelty is wrong because it causes suffering, not merely because a sacred book forbids it.

Goodness performed only to gain reward or avoid punishment is obedience.

Moral maturity asks us to care because other people matter.

9. "Atheism is also a religion or a form of faith."

Atheism, in its basic sense, is the absence of belief in gods.

It does not require scripture, worship, clergy, ritual, revelation, or belief in supernatural beings.

An atheist may hold other philosophical or political beliefs, some of which may become dogmatic. Atheists can be irrational, arrogant, or mistaken, just as believers can.

That does not make atheism itself a religion.

Declining to accept an unsupported claim is not equivalent to making the claim.

A person who does not believe in astrology has not created an anti-astrology religion.

The important question is not which label a person carries.

It is whether his beliefs remain open to evidence and correction.

10. "What harm is there in private religious belief?"

A genuinely private belief that comforts an adult and harms no one may require little public concern.

The difficulty is that religious belief rarely remains entirely private.

It influences how children are raised, which medical decisions are made, whom people marry, how women are treated, what schools teach, how money is spent, and which laws voters support.

Even a quiet belief can shape behavior.

This does not mean every religious practice is harmful. A prayer, festival, or symbolic ritual may remain personal and benign.

The relevant questions are:

Does the belief frighten anyone?

Does it claim authority over someone else?

Does it delay care?

Does it demand money?

Does it limit freedom?

Does it influence public policy without public evidence?

Belief deserves privacy.

Other people deserve protection from its consequences.

11. "Religion has done much good. Why focus on the harm?"

Religion has inspired charity, art, architecture, music, education, community service, courage, and resistance to injustice.

These contributions should be acknowledged honestly.

But good consequences do not prove supernatural claims.

Nor do they erase the harms carried out through sacred authority.

A fair evaluation must examine both.

The existence of religious hospitals does not make faith healing reliable. A beautiful temple does not prove the existence of its deity. A charitable institution does not make every doctrine humane.

People can preserve generosity, service, music, community, and moral courage without protecting false claims or harmful authority.

Goodness should be retained.

Error should still be corrected.

12. "Why not let people believe whatever makes them happy?"

Adults should generally be free to hold beliefs, including beliefs others consider mistaken.

Freedom of conscience includes the freedom to believe and the freedom not to believe.

But happiness cannot be the only standard.

A person may feel reassured by a belief that frightens his child, restricts his daughter, delays treatment, or condemns an outsider.

The believer's comfort does not cancel another person's suffering.

Freedom of belief does not include an unlimited right to impose unsupported claims on others.

A humane boundary is simple:

Believe privately as you choose.

Do not use belief to deny another person evidence, care, equality, dignity, or freedom.

A Final Standard

Religious claims should not be rejected merely because they are religious.

They should be examined by the same standards used for other serious claims.

Ask:

What exactly is being claimed?

What evidence supports it?

What evidence would weaken it?

Are failures counted?

Would the same reasoning persuade me if it supported another religion?

Who may be harmed if the claim is wrong?

A belief does not become true because it is ancient.

It does not become factual because it is comforting.

It does not become moral because it is sacred.

Respect the person.

Examine the claim.

Follow the evidence.

Appendix B: Evidence and Further Reading

Evidence is not a collection of quotations from authors who already agree with us.

It is a method.

Claims must be compared with observation, records, experiments, historical context, competing explanations, and the consequences they produce in human life.

This book has drawn from psychology, anthropology, philosophy, history, evolutionary biology, medicine, public health, human-rights scholarship, and the study of religion. No single field answers every question raised here. No single author should be treated as a new authority beyond criticism.

The purpose of this appendix is therefore not to replace sacred books with secular books that readers are expected to obey.

It is to widen the investigation.

Some works listed below are strongly critical of religion. Others explain religion sympathetically or defend religious belief. Some are written for general readers. Others are scholarly studies. Their conclusions do not always agree.

That disagreement is useful.

Readers should compare arguments, examine methods, and remain willing to revise their conclusions.

A Core Reading List

Readers seeking a manageable introduction may begin with the following works.

Carl Sagan, *The Demon-Haunted World: Science as a Candle in the Dark*

A humane and highly accessible defense of scientific thinking. Sagan explains how unsupported claims spread, why people are vulnerable to deception, and why skepticism is necessary in a free society.

Pascal Boyer, *Religion Explained: The Evolutionary Origins of Religious Thought*

An influential examination of religion as a natural product of human cognition, memory, social relationships, and cultural transmission.

Daniel C. Dennett, *Breaking the Spell: Religion as a Natural Phenomenon*

An argument that religion should be investigated openly as part of human behavior rather than protected from ordinary study because it is sacred.

Daniel Kahneman, *Thinking, Fast and Slow*

Not primarily a book about religion, but essential for understanding intuitive judgment, cognitive shortcuts, selective reasoning, and the human mind's vulnerability to error.

Carol Tavris and Elliot Aronson, *Mistakes Were Made (But Not by Me)*

A readable study of self-justification and the difficulty people experience when evidence threatens a cherished belief, identity, or past decision.

Michael Shermer, *Why People Believe Weird Things*

An accessible examination of superstition, pseudoscience, conspiracy thinking, and the mental habits that allow unsupported claims to survive.

David Hume, "Of Miracles," in *An Enquiry Concerning Human Understanding*

A classic philosophical discussion of testimony, probability, and the standard of evidence needed before accepting reports that appear to violate ordinary experience.

Jerry A. Coyne, *Why Evolution Is True*

A clear presentation of evidence for evolution from fossils, genetics, anatomy, biogeography, and observed biological change.

B. R. Ambedkar, *Annihilation of Caste*

Essential reading on inherited hierarchy, sacred authority, caste injustice, and the moral necessity of placing human dignity above tradition.

Martha C. Nussbaum, *Liberty of Conscience*

A defense of religious freedom, equality, and the principle that freedom of religion must protect minorities, dissenters, and nonbelievers.

These authors do not agree on every question. Their value lies not in their authority, but in the evidence, arguments, and methods they place before the reader.

Why Human Beings Believe

Religious belief does not arise simply because people lack education.

Human beings naturally detect patterns, imagine intention, trust familiar authorities, remember striking coincidences, and seek protection from uncertainty. Beliefs connected with family and identity receive especially strong emotional defense.

Useful works include:

- Pascal Boyer, *Religion Explained*
- Justin L. Barrett, *Why Would Anyone Believe in God?*
- Jesse Bering, *The Belief Instinct*
- Scott Atran, *In Gods We Trust*
- Ara Norenzayan, *Big Gods*
- Jonathan Haidt, *The Righteous Mind*
- Leon Festinger, Henry Riecken, and Stanley Schachter, *When Prophecy Fails*
- Carol Tavris and Elliot Aronson, *Mistakes Were Made (But Not by Me)*

Useful research articles include:

- Ziva Kunda, "The Case for Motivated Reasoning," *Psychological Bulletin*
- Jennifer A. Whitson and Adam D. Galinsky, "Lacking Control Increases Illusory Pattern Perception," *Science*
- Emily Pronin, "Perception and Misperception of Bias in Human Judgment," *Trends in Cognitive Sciences*

These works help explain why intelligent and educated people may continue protecting inherited beliefs.

They do not, by themselves, prove that no god exists. Psychological explanations of belief and the truth of a belief are separate questions.

Their importance is more limited and precise: they show that certainty, personal experience, and sincerity can arise through ordinary human processes.

Understanding this should reduce contempt.

Believers and skeptics share the same vulnerable human mind.

Fear, Ritual, and Superstition

Ritual can provide structure, community, and emotional relief. It can also create an illusion of control when events are uncertain.

Useful works include:

- Bronisław Malinowski, *Magic, Science and Religion and Other Essays*
- Stuart A. Vyse, *Believing in Magic: The Psychology of Superstition*
- Mary Douglas, *Purity and Danger*
- E. E. Evans-Pritchard, *Witchcraft, Oracles and Magic Among the Azande*
- Thomas Gilovich, *How We Know What Isn't So*
- Robert A. Hahn, *Sickness and Healing: An Anthropological Perspective*

These works examine how human beings connect unrelated events, assign meaning to ritual, and preserve beliefs even when predictions fail.

Astrology has also been examined under controlled conditions. One widely discussed study is:

- Shawn Carlson, "A Double-Blind Test of Astrology," *Nature*

No single experiment settles every version of astrology. A reliable system, however, should produce results better than chance under fair and repeatable conditions.

A claim should not be protected by becoming more vague whenever testing begins.

Miracles and Extraordinary Claims

Miracle stories often arise from experiences that are emotionally powerful:

a recovery from illness,

a prayer followed by good news,

a vivid dream,

an unusual coincidence,

a vision,

or an event that appears to resist explanation.

These experiences should not be mocked.

They should be examined.

Useful works include:

- Carl Sagan, *The Demon-Haunted World*
- David Hume, "Of Miracles"
- James Randi, *The Faith Healers*
- Joe Nickell, *Looking for a Miracle*
- Benjamin Radford, *Scientific Paranormal Investigation*

- Richard Wiseman, *Paranormality*

Studies of intercessory prayer include:

- Herbert Benson and colleagues, the Study of the Therapeutic Effects of Intercessory Prayer in cardiac bypass patients, published in the *American Heart Journal*
- Leanne Roberts, Irshad Ahmed, and Andrew Hall, "Intercessory Prayer for the Alleviation of Ill Health," *Cochrane Database of Systematic Reviews*

Such studies examine specific measurable claims. They do not test every possible meaning of prayer, such as whether prayer gives emotional comfort or helps a family endure grief.

The essential distinction remains:

an unexplained event is not automatically a supernatural event.

A serious miracle claim should welcome:

- clear documentation
- independent examination
- complete reporting of successes and failures
- comparison with ordinary outcomes
- careful consideration of natural explanations
- willingness to revise the claim if evidence does not support it

Sincerity may establish that a witness believes the story.

It does not establish that the interpretation is correct.

Science and Sacred Explanations

Religious traditions developed before modern astronomy, geology, evolutionary biology, genetics, microbiology, and neuroscience.

Their explanations naturally reflect the knowledge available to earlier societies.

Science introduced a different method. It observes, measures, compares, tests, and corrects.

Useful works include:

- Charles Darwin, *On the Origin of Species*
- Jerry A. Coyne, *Why Evolution Is True*
- Richard Dawkins, *The Greatest Show on Earth*
- National Academy of Sciences and Institute of Medicine, *Science, Evolution, and Creationism*
- Ronald L. Numbers, *The Creationists*
- Maurice A. Finocchiaro, *The Galileo Affair: A Documentary History*
- Stephen Jay Gould, *Rocks of Ages*

The history of Galileo should not be reduced to a simple story in which every scientist represented reason and every religious person opposed it. The conflict involved theology, politics, personality, scientific uncertainty, and institutional authority.

The central lesson remains important.

Evidence about nature must come from nature.

A sacred text cannot determine the movement of planets, the age of the earth, the origin of species, or the cause of disease merely because believers consider the text divine.

Scientific conclusions change because they remain open to correction.

Sacred texts do not revise themselves. Later readers must reinterpret them when evidence or moral understanding changes.

Medicine, Faith Healing, and Public Health

Religion may support patients emotionally.

Prayer, community, music, ceremony, and the presence of caring people may reduce loneliness and help someone endure illness.

These effects should not be confused with diagnosis or treatment.

Useful works include:

- Richard P. Sloan, *Blind Faith: The Unholy Alliance of Religion and Medicine*
- Harold G. Koenig, Dana E. King, and Verna Benner Carson, *Handbook of Religion and Health*
- Kenneth I. Pargament, *The Psychology of Religion and Coping*
- James Randi, *The Faith Healers*
- World Health Organization resources on mental health, epilepsy, vaccination, and harmful health practices

A particularly important study is:

- Seth M. Asser and Rita Swan, "Child Fatalities from Religion-Motivated Medical Neglect," *Pediatrics*

Readers should distinguish several different questions:

Does prayer comfort the patient?

Does belonging to a supportive community improve coping?

Does a specific religious intervention cure disease?

Does religious belief delay or prevent appropriate care?

These are not the same question and should not be answered with the same evidence.

Medical recommendations should be explainable, open to second opinion, testable against outcomes, and capable of change when better evidence appears.

Medicine is not infallible.

Physicians and medical institutions can be wrong, arrogant, unequal, or influenced by money and power. Those failures require accountability and better evidence, not a return to supernatural diagnosis.

Neither medicine nor religion should demand blind obedience.

Scripture, Interpretation, and Moral Change

Sacred books have shaped civilizations, law, literature, art, family life, and moral language.

They have also been used to defend hierarchy, punishment, slavery, caste, gender restrictions, persecution, and the silencing of unbelievers.

Understanding scripture therefore requires more than devotional reading.

Useful works include:

- James L. Kugel, *How to Read the Bible*
- Bart D. Ehrman, *Misquoting Jesus*
- Elaine Pagels, *The Gnostic Gospels*

- Karen Armstrong, *A History of God*
- Wilfred Cantwell Smith, *What Is Scripture?*
- Gavin Flood, *An Introduction to Hinduism*
- Wendy Doniger, *The Hindus: An Alternative History*
- Patrick Olivelle, translations and studies of the Upanishads and early Indian religious texts
- Gabriel Said Reynolds, *The Qur'an and the Bible*
- Eleanor Nesbitt, *Sikhism: A Very Short Introduction*

These authors differ significantly in method and sympathy toward religion.

Readers should ask:

Who wrote or compiled the passage?

When and where did it emerge?

What problem was it addressing?

How was it transmitted and translated?

Which writings were included or excluded?

How did later communities alter its meaning?

Who benefited from the interpretation?

Who carried its burden?

A text may remain culturally, historically, or spiritually important without being treated as perfect.

Caste, Gender, Purity, and Social Hierarchy

Religion does not create every form of inequality.

Political power, property, family systems, law, economics, and custom also sustain hierarchy.

Religion can, however, give inequality sacred legitimacy.

A social arrangement becomes harder to challenge when it is described as divine order, karma, purity, duty, or the natural place assigned by God.

Useful works include:

- B. R. Ambedkar, *Annihilation of Caste*
- B. R. Ambedkar, *Riddles in Hinduism*
- Uma Chakravarti, *Gendering Caste*
- Mary Douglas, *Purity and Danger*
- Martha C. Nussbaum, *Women and Human Development*
- Gail Omvedt, *Dalits and the Democratic Revolution*
- United Nations materials on discrimination against women, freedom of conscience, and equality before the law

These works help reveal the difference between describing inequality and justifying it.

When birth, gender, or normal biological processes are given sacred meaning, injustice can appear natural or morally required.

Human dignity must remain able to judge the tradition.

Witchcraft, Possession, and Sacred Accusation

Communities facing unexplained illness, death, mental disturbance, crop failure, or repeated misfortune may search for a supernatural cause.

The accusation often falls on someone vulnerable:

a widow,

an older woman,

a child,

a person with mental illness,

a social outsider,

or someone without powerful family protection.

Useful works include:

- Brian P. Levack, *The Witch-Hunt in Early Modern Europe*
- Ronald Hutton, *The Witch*
- E. E. Evans-Pritchard, *Witchcraft, Oracles and Magic Among the Azande*
- World Health Organization materials on mental illness and neurological disorders
- United Nations and human-rights reports on witchcraft accusations, ritual violence, and abuse of people believed to be possessed

Historical and cultural settings differ greatly. No single explanation fits every accusation.

The recurring danger is that fear becomes attached to a person.

Once that person is imagined as a supernatural threat, cruelty can be renamed protection.

Evidence becomes a shield for the accused.

It asks whether the accusation can be demonstrated and whether medical, psychological, environmental, or social causes have been ignored.

Freedom of Conscience

Freedom of religion is essential.

People should be able to worship, pray, gather, and practice peaceful religious customs without persecution.

The same freedom must protect disbelief.

A person should be able to question religion, change faith, reject a ritual, criticize scripture, or leave religion without losing safety, legal rights, family protection, or social standing.

Useful works and documents include:

- John Stuart Mill, *On Liberty*
- Bertrand Russell, "Why I Am Not a Christian"
- Martha C. Nussbaum, *Liberty of Conscience*
- A. C. Grayling, *The God Argument*
- Paul Kurtz's writings on secular humanism
- Universal Declaration of Human Rights, Article 18
- International Covenant on Civil and Political Rights, Article 18
- Convention on the Rights of the Child

The distinction is simple:

People deserve protection from violence and discrimination.

Ideas must remain open to criticism.

Freedom of religion should not become authority to impose religious rules on people who do not accept them.

Freedom of conscience includes freedom from religious coercion.

Indian Traditions of Reform and Rational Inquiry

Criticism of superstition, caste, religious hierarchy, and irrational belief is not foreign to India.

India possesses a long history of philosophical disagreement, skepticism, materialism, social reform, and resistance to sacred injustice.

Readers may explore:

- B. R. Ambedkar's writings on caste, religion, equality, and constitutional morality
- Debiprasad Chattopadhyaya, *Lokayata: A Study in Ancient Indian Materialism*
- Amartya Sen, *The Argumentative Indian*
- Raja Rammohun Roy's reform writings and opposition to sati
- Akshay Kumar Datta and the rationalist tradition of nineteenth-century Bengal
- Periyar E. V. Ramasamy's criticism of caste and religious authority
- Narendra Dabholkar and the modern anti-superstition movement
- Historical studies of women's reform, scientific education, caste resistance, and secular thought in India

These writers and reformers did not share one philosophy. Their language, goals, and historical circumstances differed.

What connects them is their refusal to accept an injustice merely because it was old, respected, or religiously protected.

Questioning religion is therefore not a rejection of Indian culture.

It is part of India's own intellectual history.

Serious Religious and Philosophical Counterarguments

Readers should also engage serious defenses of religious belief rather than relying only on weak popular arguments.

Useful works include:

- Alvin Plantinga, *Where the Conflict Really Lies*
- Richard Swinburne, *The Existence of God*
- Ian G. Barbour, *When Science Meets Religion*
- John Polkinghorne, *Science and Theology*
- David Bentley Hart, *The Experience of God*
- Sarvepalli Radhakrishnan, *The Hindu View of Life*

These authors do not defend the same concept of God or the same religious tradition.

Their arguments should be examined by the same standards used elsewhere:

Are the claims clear?

Does the conclusion follow from the evidence?

Could the reasoning support competing religions equally well?

Does the argument explain more than a natural account?

What evidence would weaken it?

A fair criticism of religion should address its strongest arguments, not only its weakest representatives.

How to Read These Sources

No book in this appendix should be treated as final.

While reading, ask:

- What exactly is the author claiming?
- What evidence is offered?
- Are competing explanations represented fairly?
- Is the evidence independent or mainly anecdotal?
- Are failures reported as carefully as successes?
- Does the argument rely on authority, emotion, or identity?
- What would lead the author to revise the conclusion?
- Who may benefit or suffer if the claim is accepted?

Apply these questions to religious writers, skeptical writers, scientists, historians, physicians, and to this book as well.

Reason is not loyalty to a particular author.

It is the willingness to let better evidence correct us.

Closing Note

The works listed here are guides, not masters.

Some explain belief through psychology. Some investigate miracles and pseudoscience. Others examine scripture, caste, gender, medicine, freedom, or the history of religious authority.

Their disagreements should not be feared.

Disagreement permits comparison.

Evidence can be examined.

Weak arguments can be rejected.

Stronger ones can survive scrutiny.

The question that began this book should accompany the reader through every source:

How do we know this is true?

A free mind does not search for another authority to obey.

It learns how to examine.

Thank You and Please Review the Book

Thank you for taking the time to read this book. I sincerely hope it encouraged you to think more deeply, question more honestly, and examine religion with greater clarity, courage, and compassion.

My purpose in writing the ***Religion: The Holy Poison*** series is simple but serious: to examine how religion affects reason, childhood, science, politics, freedom, equality, peace, and human dignity. These books are written for readers who value evidence over blind belief, humanity over scripture, and truth over inherited fear.

This series is not written to insult believers. It is written to question powerful ideas that have shaped human life for thousands of years. If this book made you pause, think, doubt, or see religion in a clearer way, then it has served its purpose.

If you found this book thoughtful, helpful, or worth reading, I would be truly grateful if you could leave a brief and honest review. Even a few sincere lines can make a meaningful difference. Your review helps other readers discover this series and encourages more open, rational, and evidence-based discussion about religion and its effects on human life.

Thank you again for your time, your trust, and your support.

Warm regards,
Dr. Prabhat Das

Other Books in This Series

Religion: The Holy Poison is a bold, thought-provoking, evidence-based series by Dr. Prabhat Das. It examines how religion weakens reason, feeds superstition, conditions children, controls minds, exploits fear, enters politics, justifies inequality, and turns sacred identity into division and hatred. Written in plain language and with historical examples, this series invites readers to question inherited beliefs and follow reason, evidence, and human dignity.

The 10 Books in This Series:

Book 1: Religion: The Enemy of Reason

Book 2: Religion: The Mass Delusion

Book 3: Religion: The Child's First Prison

Book 4: Religion: The Biggest Addiction

Book 5: Religion: The Business of Faith and Fear

Book 6: Religion: The Obedience Machine

Book 7: Religion: The Poisoning of Politics

Book 8: Religion: The Weapon of Hatred

Book 9: Religion: The Justification of Inequality

Book 10: Religion: The Curse on Humanity

www.ingramcontent.com/pod-product-compliance
Lightning Source LLC
LaVergne TN
LVHW011523200926
848484LV00001B/124